Recent Progress in Nanocomposites

Recent Progress in Nanocomposites

Edited by **Rich Falcon**

New York

Published by NY Research Press,
23 West, 55th Street, Suite 816,
New York, NY 10019, USA
www.nyresearchpress.com

Recent Progress in Nanocomposites
Edited by Rich Falcon

International Standard Book Number: 978-1-63238-397-6 (Hardback)

Contents

Preface

Every book is a source of knowledge and this one is no exception. The idea that led to the conceptualization of this book was the fact that the world is advancing rapidly; which makes it crucial to document the progress in every field. I am aware that a lot of data is already available, yet, there is a lot more to learn. Hence, I accepted the responsibility of editing this book and contributing my knowledge to the community.

This book brings forth the experiences of experts from different scientific spheres across the world on their encounters with various aspects of nanocomposite science and its uses. It documents latest findings and advancements in nanocomposites through study and research. The book elucidates the applications of nanocomposites in water treatment, super capacitors, anticorrosive and antistatic applications, and other such applications. It also sheds light on multipurpose nanocomposites, photonics of dielectric nanostructures and electron scattering in nanocomposite materials.

While editing this book, I had multiple visions for it. Then I finally narrowed down to make every chapter a sole standing text explaining a particular topic, so that they can be used independently. However, the umbrella subject sinews them into a common theme. This makes the book a unique platform of knowledge.

I would like to give the major credit of this book to the experts from every corner of the world, who took the time to share their expertise with us. Also, I owe the completion of this book to the never-ending support of my family, who supported me throughout the project.

Editor

New Frontiers in Mechanosynthesis: Hydroxyapatite – and Fluorapatite – Based Nanocomposite Powders

Bahman Nasiri–Tabrizi, Abbas Fahami,
Reza Ebrahimi–Kahrizsangi and Farzad Ebrahimi

Additional information is available at the end of the chapter

1. Introduction

Mechanosynthesis process is a solid state method that takes advantage of the perturbation of surface-bonded species by pressure or mechanical forces to enhance the thermodynamic and kinetic reactions between solids. Pressure can be applied via conventional milling equipment, ranging from low-energy ball mills to high-energy stirred mills. In a mill, the reactants are crushed between the balls and wall (horizontal or planetary ball mill, attritor, vibratory ball mill), or between rings or ring and wall (multi–ring media mill) (Bose et al., 2009). These processes cause the creation of defects in solids; accelerate the migration of defects in the bulk, increase the number of contacts between particles, and renew the contacts. In these circumstances, chemical interaction occurs between solids (Avvakumov et al., 2002). This procedure is one of the most important fields of solid state chemistry, namely, the mechanochemistry of inorganic substances, which is intensively developed; so that, a large number of reviews and papers published on this subject in the last decades (Silva et al., 2003; Suryanarayana, 2001; De Castro & Mitchell, 2002). The prominent features of this technique are that melting is not essential and that the products have nanostructural characteristics (Silva et al., 2003; Suryanarayana, 2001; De Castro & Mitchell, 2002). In the field of bioceramics, high efficiency of the mechanochemical process opens a new way to produce commercial amount of nanocrystalline calcium phosphate-based materials. A review of scientific research shows that the mechanosynthesis process is a potential method to synthesis of nanostructured bioceramics (Rhee, 2002; Silva et al., 2004; Suchanek et al., 2004; Tian et al., 2008; Nasiri–Tabrizi et al., 2009; Gergely et al., 2010; Wu et al. 2011; Ramesh et al., 2012).

On the other side, bioceramics play a vital role in several biomedical applications and have been expanding enormously the recent years (Adamopoulos & Papadopoulos, 2007). Among different forms of bioceramics, particular attentions have been placed to calcium phosphates-based powders, granules, dense or porous bodies, and coatings for metallic or polymeric implants due to their excellent biocompatibility and osteointegration properties (Marchi et al., 2009). It is well known that hydroxyapatite (HAp: $Ca_{10}(PO_4)_6(OH)_2$) is a major mineral component of bones and teeth (Zhou & Lee, 2011). Therefore, synthetic HAp has been extensively utilized as a bioceramic for maxillofacial applications owing to its excellent osteoconductive properties (Adamopoulos & Papadopoulos, 2007). Besides this field, in a variety of other biomedical applications calcium phosphates have been used as matrices for controlled drug release, bone cements, tooth paste additive, and dental implants (Rameshbabu et al., 2006). Nevertheless, HAp intrinsic poor mechanical properties (strength, toughness and hardness), high dissolution rate in biological system, poor corrosion resistance in an acid environment and poor chemical stability at high temperatures have restricted wider applications in load-bearing implants (Fini et al., 2003; Chen et al., 2005).

According to the literature (Jallot et al., 2005), the biological and physicochemical properties of HAp can be improved by the substitution with ions usually present in natural apatites of bone. In fact, trace ions substituted in apatites can effect on the lattice parameters, the crystallinity, the dissolution kinetics and other physical properties (Mayer & Featherstone, 2000). When OH^- groups in HAp are partially substituted by F^-, fluoride-substituted HAp (FHAp: $Ca_{10}(PO_4)_6(OH)_{2-x}F_x$) is obtained. If the substitution is completed, fluorapatite (FAp: $Ca_{10}(PO_4)_6F_2$), is formed. When fluoride consumed in optimal amounts in water and food, used topically in toothpaste, and mouth rinses, it increases tooth mineralization and bone density, reduces the risk and prevalence of dental caries, and helps to promote enamel remineralization throughout life for individuals of all ages (Palmer & Anderson, 2001). It is found that the incorporation of fluorine into HAp induced better biological response (Rameshbabu et al., 2006). On the other hand, the incorporation of bioinert ceramics and addition of appropriate amount of ductile metallic reinforcements into calcium phosphate-based materials has demonstrated significant improvement in structural features as well as mechanical properties. Therefore, improvements on structural, morphological, and mechanical properties of HAp ceramics have been tried by a number of researches (Cacciotti et al., 2009; Schneider et al., 2010; Farzadi et al., 2011; Pushpakanth et al., 2008; Rao & Kannan, 2002; Viswanath & Ravishankar, 2006; Gu et al., 2002; Ren et al., 2010). These studies have shown that such characteristics of HAp might be exceptionally strengthened by various methods such as making nanocomposites, use of different sintering techniques, and adding dopants. In the field of nanocomposites, an ideal reinforcing material for the HAp-based composites, which satisfies all of the requirements, has not yet been found. Thus, synthesis and characterization of novel nanostructured calcium phosphate-based ceramics provided the key target for current research. In most researches (Enayati–Jazi et al., 2012; Rajkumar et al., 2011; Choi et al., 2010), calcium phosphate-based nanocomposites were prepared using multiple wet techniques which ordinarily comprise of several step processes. Over the past decades, the mechanochemical synthesis has been extended for the production of a wide range of nanostructured materials (Suryanarayana, 2001), particularly for the synthesis of nanocrystalline calcium phosphate-based ceramics (Rhee, 2002; Suchanek et al., 2004; Tian et al., 2008; Nasiri–Tabrizi et al., 2009; Gergely et al., 2010; Wu et al. 2011; Ramesh et al., 2012). The advantages of this procedure remains on the fact that melting is not necessary and the powders are nanocrystalline (Silva et al., 2007).

In this chapter, a new approach to synthesis of HAp- and FAp-based nanocomposites via mechanochemical process is reported. The effect of high-energy ball milling parameters and subsequent thermal treatment on the structural and morphological features of the nanocomposites were discussed in order to propose suitable conditions for the large scale synthesis of HAp- and FAp-based nanocomposites. Powder X-ray diffraction (XRD), Fourier transform infrared (FT-IR) spectroscopy, and energy dispersive X-ray spectroscopy (EDX) techniques are used to provide evidence for the identity of the samples. Transmission electron microscopy (TEM), Field-Emission Scanning Electron Microscope (FE-SEM), and scanning electron microscopy (SEM) are also utilized to study of the morphological features of the nanocomposites. Literature reported that the size and number of balls had no significant effect on the synthesizing time and grain size of FAp ceramics, while decreasing the rotation speed or ball to powder weight ratio increased synthesizing time and the grain size of FAp (Mohammadi Zahrani & Fathi, 2009). On the other hand, our recent experimental results confirm that the chemical composition of initial materials and thermal annealing process are main parameters that affect the structural features (crystallinity degree, lattice strain, crystallite size) of the products via mechanochemical method (Nasiri-Tabrizi et al., 2009; Honarmandi et al., 2010; Ebrahimi-Kahrizsangi et al., 2010; Fahami et al., 2011; Ebrahimi-Kahrizsangi et al., 2011; Fahami et al., 2012). Consequently, the present chapter is focused on the mechanochemical synthesize of HAp- and FAp-based nanocomposites. In the first part of this chapter, an overview of recent development of ceramic-based nanocomposites in biomedical applications and mechanochemical process are provided. The other sections describe the application of these procedures in the current study. The effects of milling media and atmosphere to prepare novel nanostructured HAp-based ceramics are studied. Moreover, mechanochemical synthesis and characterization of nanostructured FAp-based bioceramics are investigated.

2. Recent developments of ceramic-based nanocomposites for biomedical applications

Over the past decades, innovations in the field of bioceramics such as alumina, zirconia, hydroxyapatite, fluorapatite, tricalcium phosphates and bioactive glasses have made significant contribution to the promotion of modern health care industry and have improved the quality of human life. Bioceramics are mainly applied as bone substitutes in biomedical applications owing to their biocompatibility, chemical stability, and high wear resistance. However, the potential of bioceramics in medical applications depends on its structural, morphological, mechanical, and biological properties in the biological environment. The first successful medical application of calcium phosphate bioceramics in humans is reported in 1920 (Kalita et al., 2007). After that the first dental application of these ceramics in animals was described in 1975 (Kalita et al., 2007). In a very short period of time, bioceramics have found various applications in replacements of hips, knees, teeth, tendons and ligaments and repair for periodontal disease, maxillofacial reconstruction, augmentation and stabilization of the jawbone and in spinal fusion (Kalita et al., 2007).

Today, many specialty ceramics and glasses have been developed for use in dentistry and medicine, e.g., dentures, glass-filled ionomer cements, eyeglasses, diagnostic instruments, chemical ware, thermometers, tissue culture flasks, fiber optics for endoscopy, and carriers for enzymes and antibodies (Hench, 1998). Among them, calcium phosphate-based bioceramics have been utilized in the field of biomedical engineering due to the range of properties that they offer, from tricalcium phosphates (α/β-TCP) being resorbable to HAp being bioactive (Ducheyne & Qiu, 1999). Hence, different phases of calcium phosphate-based bioceramics are used depending upon whether a resorbable or bioactive material is desired. The phase stability of calcium phosphate-based bioceramics depends significantly upon temperature and the presence of water, either during processing or in the use environment. It is found that at body temperature; only two calcium phosphates are stable in contact with aqueous media, such as body fluids. These stable phases are $CaHPO_4.2H_2O$ (dicalcium phosphate, brushite) and HAp at pH<4.2 and pH>4.2, respectively (Hench, 1998). At higher temperatures, other phases, such as α/β-TCP and tetracalcium phosphate ($Ca_4P_2O_9$) are present. The final microstructure of TCP will contain β or α-TCP depending on their cooling rate. Rapid cooling from sintering temperature gives rise to α-TCP phase only, whereas slow furnace cooling leads to β-TCP phase only. Any moderate cooling rate, in between these two results mixed phase of both β and α-TCP (Nath et al., 2009).

One of the primary restrictions on clinical use of bioceramics is the uncertain lifetime under the complex stress states, slow crack growth, and cyclic fatigue that result in many clinical applications. Two creative approaches to these mechanical limitations are use of bioactive ceramics as coatings, and the biologically active phase in composites. Because of the anisotropic deformation and fracture characteristics of cortical bone, which is itself a composite of compliant collagen fibrils and brittle HAp crystals, the Young's modulus varies ~ 7–25 GPa, the critical stress intensity ranges ~ 2–12 MPa.m$^{1/2}$, and the critical strain intensity increases from as low as ~ 600 J.m^{-2} to as much as 5000 J.m^{-2}, depending on orientation, age, and test condition. On the contrary, most bioceramics are much stiffer than bone and many exhibit poor fracture toughness (Hench, 1998). Therefore, the only materials that exhibit a range of properties equivalent to bone are composites. For this reason, many attempts have been made to improve the mechanical properties as well as structural features through the incorporation of ceramic second phases (Viswanath & Ravishankar, 2006; Evis, 2007; Nath et al., 2009; Ben Ayed & Bouaziz, 2008). These studies have shown that the mechanical properties of HAp and fluoridated HAp might be exceptionally strengthened by composite making technique.

It is found that (Kong et al., 1999) the following conditions should be satisfied to be effective as a reinforcing agent for a ceramic matrix composite material. First, the strength and the elastic modulus of the second phase must be higher than those of the matrix. Second, the interfacial strength between the matrix and the second phase should be neither too weak nor too strong. Indeed, for an appropriate interfacial strength, no excessive reaction should occur between the matrix and the second phase. Third, the coefficient of thermal expansion (CTE) of the second phase should not differ too much from that of the matrix in order to prevent micro-cracks formation in densification process. Fourth, in the case of biomaterials, the biocompatibility of the reinforcing agent is another crucial factor that should be consid- ered. Nevertheless, an ideal reinforcing material for the calcium phosphate-based compo- sites, which satisfies all of requirements, has not yet been found. So, some attempts have been made to develop HAp- and fluorhydroxyapatite-based composites such as HAp–Al_2O_3 (Viswanath & Ravishankar, 2006), HAp–ZrO_2 (Evis, 2007), HAp–TiO_2 (Nath et al., 2009), FHAp–Al_2O_3 (Adolfsson et al., 1999), FHAp–ZrO_2 (Ben Ayed & Bouaziz, 2008), poly(lactide- co-glycolide)/β-TCP (Jin. et al., 2010), polyglycolic acid (PGA)/β-TCP (Cao & Kuboyama, 2010), and HAp–CNT (Lee et al., 2011) composites. These experimental studies exhibited that interfacial reactions occurred during the high temperature processing of composites due to the large interfacial area available for the reactions. Interfacial reactions result in the formation of new phases, influence densification, mechanical properties and even degrade the biological properties of the composite in some cases which often limit their performance (Viswanath & Ravishankar, 2006). Hence, control over nanocomposite characteristics is a challenging task.

3. Mechanosynthesis of ceramic-based nanocomposites

To date, several approaches, including wet chemical methods (Mobasherpour et al., 2007; Kiv- rak & Tas, 1998), hydrothermal processes (Liu et al., 2006), solid–state reaction (Silva et al., 2003), and sol–gel method (Balamurugan et al., 2002), have been developed for synthesis of nanobioceramics. Among them, mechanochemical process has been extended for the produc- tion of a wide range of nanostructured materials (Suryanarayana, 2001; De Castro & Mitchell, 2002). According to literature (Bose et al., 2009), mechanochemical synthesis was originally de- signed for the production of oxide dispersion-strengthened (ODS) alloys. Over the past 20 years, however, the number of available mechanochemical synthesis has grown, such that Nowadays it is used for the fabrication of a wide range of advanced materials, both metallic and nonmetallic in composition. In mechanosynthesis, the chemical precursors typically con- sist of mixtures of oxides, chlorides and/or metals that react either during milling or during subsequent thermal treatment to form a composite powder consisting of the dispersion of ul- trafine particles within a soluble salt matrix. The ultrafine particle is then recovered by selec- tive removal of the matrix phase through washing with an appropriate solvent.

Mechanochemical approach is a very effective process for synthesizing nanocomposites with various classes of compounds: metals, oxides, salts, organic compounds in various combinations. For example, Khaghani-Dehaghani et al. (Khaghani-Dehaghani et al., 2011) synthesized Al_2O_3–TiB_2 nanocomposite by mechanochemical reaction between titanium di- oxide, acid boric and pure aluminum according to the following reactions:

$$2H_3BO_3 \rightarrow B_2O_3 + 3H_2O \tag{1}$$

$$3TiO_2 + 3B_2O_3 + 10Al \rightarrow 3TiB_2 + 5Al_2O_3 \tag{2}$$

Titanium diboride has an attractive combination of high Vickers hardness, electrical conductivity, excellent chemical resistance to molten nonferrous metals and relatively low specific gravity (Gu et al., 2008). However, titanium diboride has poor fracture toughness and impact strength. Thus, the composites of TiB_2 such as Al_2O_3–TiB_2 improve those mechanical properties. These nanocomposites are useful in variety of applications such as cutting tools, wear-resistant substrates, and lightweight armor (Mishra et al., 2006). Results reveal that the Al_2O_3–TiB_2 nanocomposite was successfully synthesized after 1.5 h of milling. Also, the determined amounts of structural features demonstrate that after 20 h of milling the steady state was obtained. Increasing milling time up to 40 h had no significant effect other than refining the crystallite size. The SEM and TEM observations show that increase of milling time was associated with decrease of powder particles, so that a fine structure was produced after 40 h of milling. Figure 1 shows the morphological features of the Al_2O_3–TiB_2 nanocomposite powders after 40 h of milling by SEM and TEM. It is clear that the particles exhibited high affinity to agglomerate. The agglomerates include fine particles of TiB_2 and Al_2O_3.

Thermodynamic studies, based on thermodynamic databases, show that the change in Gibbs free energy of the reduction of boron oxide and titanium oxide with aluminum (Eqs. (3) and (4)) is favorable at room temperature.

$$4Al + 3TiO_2 \rightarrow 2Al_2O_3 + 3Ti$$
$$\Delta G^\circ_{298K} = -495.488 \text{ kJ}, \quad \Delta H^\circ_{298K} = -516.306 \text{ kJ} \tag{3}$$

$$2Al + B_2O_3 \rightarrow Al_2O_3 + 2B$$
$$\Delta G^\circ_{298K} = -389.053 \text{ kJ}, \quad \Delta H^\circ_{298K} = -403.338 \text{ kJ} \tag{4}$$

It is well known if a reaction is highly exothermic, the impact of the milling balls can initiate a mechanically induced self-sustaining reaction (MSR) (Xia et al., 2008). MSR was usually observed in highly exothermic reactions. The ignition of MSR takes place after a certain activation time, during which the powder mixtures reach a critical state due to the physical and chemical changes caused by ball milling (Takacs, 2002; Takacs et al., 2006). That certain activation time depends mainly on the exothermicity of the process, the milling conditions and the mechanical properties of the raw materials. Takacs (Takacs, 2002) showed that a reaction can propagate in the form of a self sustaining process, if $\Delta H/C$, the magnitude of the heat of reaction divided by the room temperature heat capacity of the products, is higher than about 2000 K. The calculations on the system Al–B_2O_3–TiO_2 show that the value of $\Delta H/C$ is about 5110 K. Therefore, the proposed reactions occurred through an expanded MSR reaction in milled samples which led to the formation of Al_2O_3–TiB_2 nanocomposite after short milling times.

Figure 1. a) SEM micrograph and (b) TEM image of Al_2O_3–TiB_2 nanocomposite after 40 h of milling (Khaghani-Deha-ghani et al., 2011).

3.1. Mechanochemical synthesize of hydroxyapatite nanostructures

HAp and its isomorphous modifications are valuable and prospective materials in biomedi-cal applications. Therefore, a large number of studies was performed on this subject in the last decade (Rhee, 2002; Silva et al., 2004; Suchanek et al., 2004; Tian et al., 2008; Nasiri-Tabri-zi et al., 2009; Gergely et al., 2010; Wu et al. 2011; Ramesh et al., 2012). Generally, the fabrica-tion methods of HAp nanostructures can be classified into two groups: wet and dry (Rhee, 2002). The advantage of the wet process is that the by-product is almost water and as a re-sult the probability of contamination during the process is very low. On the other hand, the dry process has benefit of high reproducibility and low processing cost in spite of the risk of contamination during milling. Furthermore, the dry mechanochemical synthesis of HAp presents the advantage that melting is not necessary and the powder obtained is nanocrys-talline. The calcium and phosphorous compounds used as the starting materials in the dry process are dicalcium phosphate anhydrous ($CaHPO_4$), dicalcium phosphate dihydrate ($CaHPO_4.2H_2O$), monocalcium phosphate monohydrate ($Ca(H_2PO_4)_2.H_2O$), calcium pyro-phosphate ($Ca_2P_2O_7$), calcium carbonate ($CaCO_3$), calcium oxide (CaO), and calcium hydrox-ide ($Ca(OH)_2$), etc.

Otsuka et al. (Otsuka et al., 1994) investigated the effect of environmental conditions on the crystalline transformation of metastable calcium phosphates during grinding. Based on the results, the mixture of $CaHPO_4$ and $Ca(OH)_2$ transformed into low-crystallinity HAp after grinding in air. Nevertheless, under N_2 atmosphere, a mixture of initial materials did not transform into HAp. After that, Toriyama et al. (Toriyama et al., 1996) proposed a method to prepare powders and composite ceramic bodies with a matrix comprising HAp. The pow-ders Was produced by the utilization of a simple and economic mechanochemical method. The composite ceramic bodies were easily obtained by simple firing of the powders at a suit-able temperature (1250 C). After sintering, the obtained products exhibited a flexural strength of more than 100 MPa in standard samples. This value is significantly higher than that usually attainable with commercially available powders (60 MPa). In another research

(Yeong et al., 2001), nanocrystalline HAp phase has been produced by high-energy mechanical activation in a dry powder mixture of CaO and $CaHPO_4$. The initial stage of mechanical activation resulted in a significant refinement in crystallite and particle sizes, together with a degree of amorphization in the starting powder mixture. A single-phase HAp of high crystallinity was attained by >20 h of mechanical activation. The resulting HAp powder exhibits an average particle size of ~ 25 nm. It was sintered to a density of 98.20% theoretical density at 1200 C for 2 h. The hardness increases almost linearly with rising sintering temperature from 900 to 1200 C, where it reaches a maximum of 5.12 GPa. This is followed by a slight decrease, to 4.92 GPa, when the sintering temperature is raised to 1300 C. Afterward, Rhee (Rhee, 2002) synthesized HAp powder by mechanochemical reaction between $Ca_2P_2O_7$ and $CaCO_3$. The two powders were mixed in acetone and water, respectively, and the single phase of HAp was observed to occur only in the powder milled in water, without the additional supply of water vapor during heat-treatment at 1100 C for 1 h. The results indicated that the mechanochemical reaction could supply enough amount of hydroxyl group to the starting powders to form a single phase of HAp. Therefore, the powder of high crystalline HAp can be obtained by the simple milling in water and subsequent heat-treatment. With the development of nanostructured materials using mechanochemical processes, nanocrystalline powders of HAp was produced in 2003 by Silva et al. (Silva et al., 2003). To produce nanocrystalline powders of HAp, five different experimental procedures in a pure dry process were utilized. For four different procedures, HAp was obtained after a couple of hours of milling (in average 60 h of milling, depending in the reaction procedure). In the preparation of nanocrystalline HAp, commercial oxides $Ca_3(PO_4)_2 \cdot xH_2O$, $Ca(OH)_2$, $CaHPO_4$, P_2O_5, $CaCO_3$ and $(NH_4)H_2PO_4$ were used in the HAp preparation. This milling process, presents the advantage that melting is not necessary and the powder obtained is nanocrystalline with crystallite size in the range of 22 nm to 39 nm. Subsequently, Silva et al. (Silva et al., 2004) synthesized nanocrystalline powders of HAp using three different experimental procedures (HAPA: $Ca(H_2PO_4)_2 + Ca(OH)_2$; HAPB: $Ca(H_2PO_4)_2 + CaCO_3$; and HAPC: $CaHPO_4 + CaCO_3$). Nanocrystalline HAp was obtained after 5, 10 and 15 h of milling in the reactions HAPA and HAPB, but it is necessary 15 h of milling in the reaction HAPC to obtain HAP. Moreover, in order to improve the mechanical properties of HAp calcium phosphate ceramics, with titanium (CaP-Ti) and zirconium (CaP-Zr), were prepared by dry ball milling using two different experimental procedures: CaP-Ti1: $Ca(H_2PO_4)_2 + TiO_2$; CaP-Ti2: $CaHPO_4 + TiO_2$; and CaP-Zr1: $Ca(H_2PO_4)_2 + ZrO_2$, CaP-Zr2: $CaHPO_4 + ZrO_2$. The calcium titanium phosphate phase, $CaTi_4P_6O_{24}$, was produced in the reaction CaP-Ti1. In the reactions CaP-Ti2, CaP-Zr1 and CaP-Zr2, it was not observed the formation of any calcium phosphate phase even after 15 h of dry mechanical alloying.

Nanocrystalline HAp powders were synthesized by the mechanochemical–hydrothermal method using emulsion systems consisting of aqueous phase, petroleum ether (PE) as the oil phase and biodegradable Tomadol 23–6.5 as the nonionic surfactant (Chen et al., 2004). $(NH_4)_2HPO_4$ and $Ca(NO_3)_2$ or $Ca(OH)_2$ were used as the phosphorus and calcium sources, respectively. The calcium source and emulsion composition had significant effects on the stoichiometry, crystallinity, thermal stability, particle size, and morphology of final products.

Disperse HAp crystals with a 160 nm length were formed in an emulsion system containing 10 wt% PE, 60 wt% water, and 30 wt% surfactant. The HAp particles had needle morphology with a specific surface area of 190 m^2/g. According to obtained results, HAp nanopowders with specific surface areas in the range of 72–231 m^2/g were produced. In the same year, Mochales et al. (Mochales et al., 2004) investigated the possibility of mechanochemistry to synthesize calcium deficient HAp (CDHA) with an expected molar calcium to phosphate (Ca/P) ratio ± 0.01. To optimize the experimental conditions of CDHA preparation from dicalcium phosphate dihydrate (DCPD) and calcium oxide by dry mechanosynthesis reaction, the kinetic study was carried out with two different planetary ball mills (Retsch or Fritsch Instuments). Results obtained with the two mills led to the same conclusions although the values of the rate constants of DCPD disappearance and times for complete reaction were very different. Certainly, the origin of these differences was from the mills used, thus the influence of instrumental parameters such as the mass and the surface area of the balls or the rotation velocity were examined on the mechanochemical reaction kinetics of DCPD with CaO. Results exhibited that the DCPD reaction rate constant and the inverse of the time for complete disappearance of CaO both vary linearly with (i) the square of the rotation velocity, (ii) the square of eccentricity of the vial on the rotating disc and (iii) the product of the mass by the surface area of the balls. The consideration of these four parameters allows the transposition of experimental conditions from one mill to another or the comparison between results obtained with different planetary ball mills. Gonzalez et al. (Gonzalez et al., 2006) studied the mechanochemical transformation of two mixtures: $Ca(OH)_2$–$(NH_4)_2HPO_4$ and $Ca(OH)_2$–P_2O_5, milled in a mortar dry grinder for different periods of time. Mechanical grinding and thermal treatment was a successful method to obtained biphasic mixtures of HAp/β-TCP. Amorphization, for both reactant mixtures, was observed after prolonged milling, 17.5 h for $Ca(OH)_2$–$(NH_4)_2HPO_4$ mixture and 5 h for the $Ca(OH)_2$–P_2O_5 mixture. The composition of the milled powders varied in the range of 1.50 < Ca/P < 1.67 for different milling periods. Calcination of milled powders of both mixtures at 800 C led to the formation of HAp and β-TCP, with an average particle size of 200 nm. Further, the Ca/P ratio affects the proportion of HAp and β-TCP phases obtained after thermal treatment. Also, Kano et al. (Kano et al., 2006) developed a novel mechanochemical process to prepare HAp fine particles. For this aim, a non-thermal process for dechlorinating of Polyvinyl chloride (PVC) was utilized. This process was composed of two steps: The first step was to grind the PVC waste with an active grinding additive such as CaO, leading to transformation of organic chlorine into water soluble chloride mechanochemically. The second step is to remove the formed chloride from the milled product by washing with water. When the filtrate was mixed with solution which contains phosphate ion $PO_4{}^{2-}$, HAp fine particles formed which has sorption ability for heavy metals such as Pb^{2+}. El Briak-BenAbdeslam et al. (El Briak-BenAbdeslam et al., 2008) investigated the influence of water addition on the kinetics of the mechanochemical reaction of dicalcium phosphate dihydrate with calcium oxide. The DCPD disappearance rate constant k and the final reaction time t_f were determined in each case and correlated with the water content present in the slurry. Results showed that the addition water (i) slowed down the reaction rate and (ii) increased the powder contamination by mill material (hard porcelain) due to ball and vial erosion; and that (iii) wet milling did not generate the expected products, in contrast to dry grinding, because porcelain induced HAp decomposition with the formation of β-TCP and silicon-stabilized tricalcium phosphate. Consequently, dry mechanosynthesis appears preferable to wet milling in the preparation of calcium phosphates of biological interest.

3.1.1. Single-crystal hydroxyapatite nanoparticles

A new approach to mechanochemical synthesis of HAp nanostructures was developed in 2009 by Nasiri-Tabrizi et al. (Nasiri–Tabrizi et al., 2009). Single-crystal HAp nanorods and nanogranules synthesized successfully by a mechanochemical process using two distinct experimental procedures.

$$6CaHPO_4 + 4Ca(OH)_2 \rightarrow Ca_{10}(PO_4)_6(OH)_2 + 6H_2O \tag{5}$$

$$4CaCO_3 + 6CaHPO_4 \rightarrow Ca_{10}(PO_4)_6(OH)_2 + 4CO_2 + 2H_2O \tag{6}$$

The feasibility of using polymeric milling media to prepare HAp nanoparticles is described. By controlling the temperature and milling time during mechanical activation (45-min milling steps with 15-min pauses), powders with three different crystallite size, lattice strain and crystallinity degrees are produced. Figure 2 presents the XRD patterns of reactions 5 and 6, respectively. The XRD patterns show that the product of reaction 5 is HAp. The extra peaks ($CaHPO_4$, ■) occurred in $2\theta = 26.59$ and 30.19 , consecutively. In reaction 6, the extra peaks are not observed after 40, 60 and 80 h of milling and the only detected phase is HAp, as shown in Figure 2(b). Therefore, during milling process, $CaHPO_4$ is a compound that should be avoided if the purpose is to achieve pure HAp without any extra phase presentation. In order to determine crystallite size and lattice strain in activated samples, the full width at half maximum (FWHM) of each peak is usually considered. Furthermore, the fraction of crystalline phase (Xc) in the HAp powders is evaluated by Landi equation (Landi et al., 2000).

According to obtained data, the crystallite size decreases and the lattice strain increases with increase of milling time. However, the rate of both variations, i.e. increasing lattice strain and decreasing crystallite size, decreases by increasing the milling time. Furthermore, the obtained data show that by choosing the total milling time to 80 h for reaction 5, the crystallinity degree increases first and reaches to a maximum at 60 h of milling, and then by further increasing the milling time to 80 h, the crystallinity degree decreases. Moreover, the increase of HAp crystallinity compared to the increase of milling time was not linear. The fraction of crystalline phase in the HAp powders from reaction 6 indicates that by increasing the milling time from 40 to 80 h, the crystallinity degree decreases mostly after 60 h and reaches to a minimum at 80 h of milling time. Based on these results, we conclude that the chemical composition of initial materials and the milling time are important parameters that affect the structural properties of product via mechanochemical process.

The morphological features of the synthesized HAp products were further examined by TEM technique. Figures 3 and 4 show the TEM micrographs of nanorods and nanogranules, respectively. Figure 3a shows that the sample possesses a mostly rod-like structure after 60 h milling time in polymeric milling vial for reaction 5. In Figure 3b, it can be seen that the morphology of nanocrystalline HAp after 80 h milling time, similar to 60 h, is also the rod shape; although, few particles appear to be close to a spherical shape. Using HAp nanorods as raw materials is an effective way to obtain dense bioceramics with high mechanical properties. Hence, this product may be used as strength enhancing additives for the preparation of the HAp ceramics or biocompatible nanocomposites.

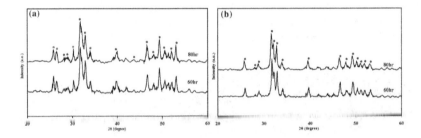

Figure 2. XRD patterns of samples milled for 60 and 80 h, (a) reaction 5 and (b) reaction 6. (Nasiri-Tabrizi et al., 2009).

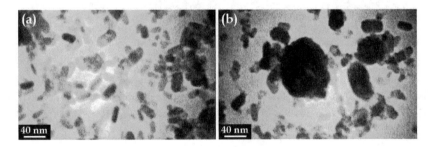

Figure 3. Typical TEM micrograph of nanorods HAp after 60 h (a) and 80 h (b) milling time for reaction 5 (Nasiri-Tabrizi et al., 2009).

In reaction 5, more agglomeration also occurs by increasing milling time from 60 h to 80 h. In fact, the obtained product nearly had a uniform geometry distribution just after 60 h milling time. Although, it may appear some ellipse or round like shapes from this image, it is due to the axis orientation of nanorods with respect to the image plane. In other words, if the rod axis is perpendicular or oblique on the image plane, the rod may be seen as a full circle or ellipse, respectively. Despite of previous research that a perfect spherical shape rarely observed in the mechanically alloyed powders, nanosphere particles were successfully obtained. In Figure 4, it can be seen that the morphology of nanocrystalline HAp for reaction 6, either after 60 or 80 h milling time, is absolutely spherical granules with a reasonable smooth geometry.

Therefore, we reach to an important conclusion that using polyamide-6 milling vial leads to the spherical granules HAp. Since spherical geometry compared to irregular shape is important for achieving osseointegration (Komlev et al., 2001; Nayar et al., 2006; Hsu et al., 2007), the latest product is well preferred for medical applications. Similar to previous reaction, the obtained product after 60 h has a better uniform geometry distribution than one after 80 h milling time. It should be noted that the HAp particles out of reaction 5 are in average length of 17 ± 8 nm and 13 ± 7 nm after 60 and 80 h milling time, respectively. Similarly, the HAp particles out of reaction 6 are in average diameter of 16 ± 9 nm and 15 ± 8 nm after 60 and 80 h milling time. Based on obtained data, the maximum particle distribution is below the crystallite size which is estimated from the line broadening of the given X-ray diffraction peak.

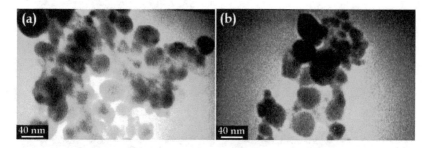

Figure 4. Typical TEM micrograph of nanospheres HAp after 60 h (a) and 80 h (b) milling time for reaction 6 (Nasiri-Tabrizi et al., 2009).

Thus, after 80 h milling time, we ascertain that this method gives rise to the single-crystal HAp with their average size below 20 nm and 23 nm for reactions 6 and 7, respectively. In fact, a novel method for the synthesis of nanosize single-crystal HAp is developed in both spherical and rod-like particles.

3.1.2. Milling media effects on structural features of hydroxyapatite

Honarmandi et al. (Honarmandi et al., 2010) investigated the effects of milling media on synthesis, morphology and structural characteristics of single-crystal HAp nanoparticles. Typical TEM images of nanosize HAp particles produced through reactions 5 and 6 after being milled in both metallic and polymeric vials have been shown in Figure 5.

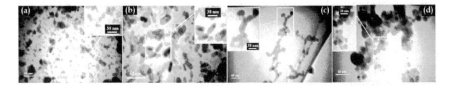

Figure 5. Morphologies of HAp synthesized through reactions 5 after being milled for 60 h in (a) metallic vials and (b) polymeric vials; through reactions 6 after being milled for 60 h in (c) metallic vials and (d) polymeric vials.

The results reveal that the single-crystal HAp nanoparticles have been successfully produced in metallic and polymeric vials through two different experimental procedures. Transmission electron microscopy images illustrate the wide morphology spectrums of the single-crystal HAp nanoparticles which are ellipse-, rod- and spherical-like morphologies each of which can be applied for specific purpose. After 60 h milling, this method results in the single-crystal HAp with their average sizes below 21 and 24 nm in the tempered chrome steel and polyamide-6 vials, respectively. According to TEM images the obtained single-crystal HAp in polymeric vials have more production efficiency and better uniform geometry distribution than products in metallic vials. In metallic vial, intense agglomeration happens during mechanochemical process as shown in Figure 6. Therefore, an important conclusion reaches that the polyamide-6 vial is more suitable than the tempered chrome steel vial for the synthesis of single-crystal HAp nanoparticles with appropriate morphology.

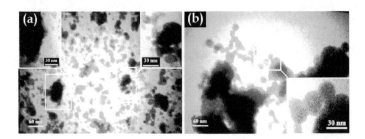

Figure 6. TEM images of agglomerated products which is obtained after 60 h milling in metallic vials through a (a) reaction 5 and (b) reaction 6 (Honarmandi et al., 2009).

3.1.3. Milling atmosphere effect on structural features of hydroxyapatite

In recent years, various mechanochemical processes were utilized to synthesis HAp nanostructures. For instance, Gergely et al. (Gergely et al., 2010) synthesized HAp by using recycled eggshell. The observed phases of the synthesized materials were dependent on the mechanochemical activation method (ball milling and attrition milling). Attrition milling proved to be more efficient than ball milling, as resulted nanosize, homogenous HAp even after milling. SEM micrographs showed that the ball milling process resulted in micrometer sized coagulated coarse grains with smooth surface, whereas attrition milled samples were characterized by the nanometer size grains. Wu et al. (Wu et al. 2011) synthesized HAp from oyster shell powders by ball milling and heat treatment. The wide availability and the low cost of oyster shells, along with their biological– natural origin are highly attractive properties in the preparation of HAp powders for biomedical application. Chemical and microstructural analysis has shown that oyster shells are predominantly composed of calcium carbonate with rare impurities. Solid state reactions between oyster shell powders (calcite polymorph of $CaCO_3$) and calcium pyrophosphate ($Ca_2P_2O_7$) or dicalcium phosphate dihydrate ($CaHPO_4.2H_2O$, DCPD) were performed through ball milling and subsequently heat treatment. The ball milling and heat treatment of $Ca_2P_2O_7$ and oyster shell powders in air atmosphere produced mainly HAp with a small quantity of β-TCP as a by product. However, oyster shell powder mixed with DCPD and milled for 5 h followed by heat-treatment at 1000 C for 1 h resulted in pure HAp, retaining none of the original materials.

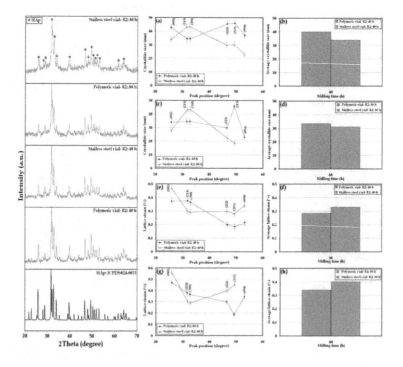

Figure 7. XRD patterns, crystallite size, lattice strain and their average of samples milled for 40 and 80 h in polymeric and metallic vials under argon atmosphere.

Mechanosynthesis of calcium phosphates can be performed under air or inert gas atmosphere. In most papers and patents, grinding under air atmosphere was selected. So far, only a few papers were devoted to mechanosynthesis of calcium phosphates under inert gas atmosphere (Nakano et al, 2001). To understand the effect of inert gas atmosphere, the mechanochemical synthesis under argon atmosphere was investigated by our research group. The starting reactant materials are $CaCO_3$ and $CaHPO_4$. The initial powders with the desired stoichiometric proportionality were mixed under a purified argon atmosphere (purity> 99.998 vol %). Figure 7 shows the XRD patterns of the powder mixture after 40 and 80 h of milling in the polymeric and metallic vials under argon atmosphere. The XRD patterns of obtained powders exhibit that the production of mechanical activation is single phase HAp. Also, Figure 7 illustrates the determined amounts of crystallite size; lattice strain and their average for experimental outcomes after 40 and 80 h of milling in polymeric and metallic vials under argon atmosphere.

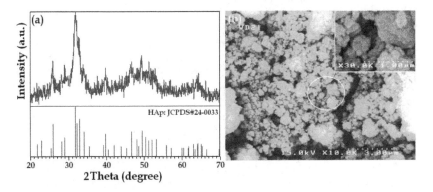

Figure 8. Typical TEM micrograph of nanocrystalline HAp after 80 h of milling under argon atmosphere in (a) polymeric and (b) metallic vials.

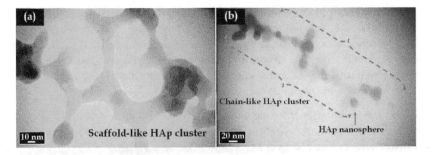

Figure 9. a) XRD profile and (b) FE-SEM images of nanocrystalline HAp with low degree of crystallinity after 2 h of milling in polymeric vial under air atmosphere.

Using the (0 0 2) plane (Figure 7a), the crystallite size of HAp is around 43 and 34 nm after 40 h of milling in polymeric and metallic vials, respectively. For comparison, the mean values determined from the use of six planes simultaneously, i.e. (0 0 2), (2 1 1), (3 0 0), (2 2 2), (2 1 3), and (0 0 4) planes. The calculated data indicates that the average crystallite size of HAp is around 40 and 34 nm, respectively. Moreover, using the (0 0 2) plane the crystallite size of HAp is around 34 and 28 nm after 80 h of milling in polymeric and metallic vials, respectively. However, the average crystallite size of HAp is around 34 and 31 nm after 80 h of milling in polymeric and metallic vials, respectively. The evaluation of the lattice strain of HAp reveals that the average of lattice strain partially increased from 0.286 % to 0.340 % after 80 h of milling in polymeric vial. A similar trend was observed in the average lattice strain of HAp after 80 h of milling. According to Figure 7, the average crystallite size decreases and the average lattice strain increases with increase of milling time from 40 up to 80 h. The TEM micrographs of synthesized powder after 80 h of milling in polymeric and metallic vials under argon atmosphere are shown in Figure 8. The TEM micrographs show that HAp particles can attach at crystallographically specific surfaces and form scaffold- and chain-like cluster composed of many primary nanospheres. In is found that (Pan et al., 2008) the living organisms build the outer surface of enamel by an oriented assembly of the rod-like crystal and such a biological construction can confer on enamel protections against erosion. It should be noted that, comparison of the physical, mechanical and biocompatibility between classical HAp ceramics and the novel nanostructures will be carried out in our laboratory.

Whilst the main advantages of the mechanochemical synthesis of ceramic powders are simplicity and low cost, the main disadvantages are the low crystallinity and calcium-deficient nonstoichiometry (Ca/P molar ratio 1.50 – 1.64) of the HAp powders, as this results in their partial or total transformation into β-TCP during calcination (Bose et al., 2009). Hence, control over crystallinity degree of HAp nanostructures for specific applications is a challenging task. Based on experimental results, we conclude that the chemical composition of initial materials, milling time, milling media, and atmosphere are important parameters that affect the structural properties (crystallite size, lattice strain, crystallinity degree) and morphological features of HAp nanostructures during mechanochemical process. For example, mechanical activation of $Ca(OH)_2$ and P_2O_5 powder mixture lead to the formation of single phase HAp with low fraction of crystallinity (Figure 9). According to this mechanochemical reaction (7), nanocrystalline HAp with an average crystallite size of about 14 nm was produced after 2 h of milling in polymeric vial under air atmosphere. In addition the fraction of crystallinity was around 7 %.

$$10Ca(OH)_2 + 3P_2O_5 \rightarrow Ca_{10}(PO_4)_6(OH)_2 + 9H_2O \tag{7}$$

Figure 9b shows the morphology and particle size distribution of the nanocrystalline HAp produced after 2 h of milling. From the FE–SEM micrograph, it is clear that the powders displayed an agglomerate structure which consisted of several small particles with the average size of about 58 nm. In the field of science and technology of particles, agglomerate size is one of the key factors that influence the densification behaviors of nanoparticles. Large particle size along with hard agglomerates shows lower densification in calcium phosphate ceramics due to the formation of large interagglomerate/intraagglomerate pores (Banerjee et al., 2007). The large interagglomerate/intraagglomerate pores increase the diffusion distance, resulting in lowering the densification rate. Thus, to compensate for this, higher sintering temperature becomes necessary.

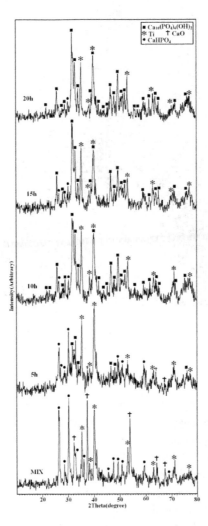

Figure 10. XRD patterns of the HAp-20%wt Ti nanocomposite after mechanochemical process for various time periods (Fahami et al., 2011).

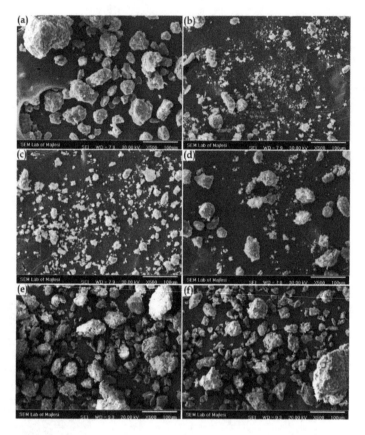

Figure 11. SEM micrographs of the HAp-20 wt.% Ti nanocomposite after different milling times (a) 5, (b) 10, (c) 15, (d) 20, (e) 40, and (f) 50 h.

3.1.4. Hydroxyapatite/titanium (HAp-Ti) nanocomposite

Apart from the displacement reactions to reduce oxides, chlorides, and sulfides to pure metals, mechanical alloying technique was also used to synthesize a large number of commercially important alloys, compounds, and nanocomposites using the mechanochemical reactions (Suryanarayana, 2001; De Castro & Mitchell, 2002; Balaz, 2008). An important characteristic of mechanosynthesized composites is that they have nanocrystalline structures which could improve the mechanical as well as biological properties (Silva et al., 2007). Nowadays, ceramic nanocomposites which play a crucial role in technology can be synthesized using surprisingly simple and inexpensive techniques such as a mechanochemical method which ordinarily include a two step process. Considering the above characteristics of the ceramic-based composites, the possibility of using one step mechanochemical process as a simple, efficient, and inexpensive method to prepare HAp-20wt.% Ti nanocomposite was investigated by our research group (Fahami et al., 2011). Furthermore, crystallite size, lattice strain, crystallinity degree, and morphological properties of products were determined due to the biological behaviour of HAp ceramics depends on structural and morphological features. For the preparation of HAp-20 wt.% Ti nanocomposite, anhydrous calcium hydrogen phosphate and calcium oxide mixture with Ca/P = 1.67 ratio was milled with the distinct amount of elemental titanium (20 wt.%) during 0, 5, 10, 15, and 20 h by a high energy planetary ball mill under highly purified argon gas atmosphere. The following reaction can be occurred at this condition (8):

$$6CaHPO_4 + 4CaO + Ti \rightarrow Ca_{10}(PO_4)_6(OH)_2 + Ti + 2H_2O \tag{8}$$

Figure 10 shows the XRD patterns of the samples after mechanochemical process for various time periods. At the initial mixture, only sharp characteristic peaks of $CaHPO_4$, CaO and Ti are observed. With increasing milling time to 5 h, the sharp peaks of starting materials degraded significantly, but the decreasing rate of each initial powder was differed. On the other hand, the appearance of weak peak between 31 and 32 confirms the formation of HAp phase. The main products of powder mixtures after 10 h of milling were HAp and Ti. The XRD patterns of the samples which are milled for 15 and 20 h indicate that increasing milling time to above 10 h does not accompany with any phase transformation. The determined amounts of crystallite size and lattice strain of the samples, after different milling time were presented in Table 1. According to Table 1, the crystallite size of HAp decrease with increasing milling time up to 20 h; whereas the change in crystallite size of Ti with increasing milling time is not linear. The calculated amount of crystallinity degree indicate that the increasing milling time dose not accompany by remarkable change in degree of crystallinity. Since the amorphous powders could find applications to promote osseointegration or as a coating to promote bone ingrowth into prosthetic implants (Sanosh et al., 2009), the resultant powders could be used to various biomedical applications.

Samples	Milling time (h)	Crystallite size (nm)		Lattice strain (%)		Crystallinity (%)
		HAp	Ti	HAp	Ti	
I	5	-	22.8	-	0.494	-
II	10	20.13	18.7	0.492	0.621	17
III	15	19.06	19.3	0.522	0.601	15
IV	20	13.13	22	1.209	0.530	13

Table 1. Comparison between structural features of the samples after different milling times.

The SEM micrographs of the samples after different milling times are presented in Figure 11. It can be seen that the particles of products can be attached together at specific surfaces and form elongated agglomerates which composed of many primary crystallites. The agglomerates with flaky-like structure formed after 10 h of milling. It seems that the existence of ductile Ti can be led to the more agglomeration during mechanochemical process. With increasing milling time to 20 h owing to sever mechanical deformation introduced into the powder, particle, and crystal refinement have occurred. Based on SEM observations, milling process reached steady state after 40 h of milling where the particles have become homogenized in size and shape. Figure 12 shows the SEM images of the HAp-20wt.% Ti nanocomposite after 40 and 50 h of milling and subsequent heat treatment at 700 C for 2h. According to SEM observations, the annealing of the milled samples at 700 C demonstrates the occurrence of grain growth.

Figure 12. SEM micrographs of the HAp-20 wt.% Ti nanocomposite after different milling times (a) 40 and (b) 50 h and subsequent heat treatment at 700 C for 2h.

3.1.5. Hydroxyapatite/geikielite (HAp/MgTiO₃–MgO) nanocomposite

In the field of nanocomposites, an ideal reinforcing material for calcium phosphate-based composites has not yet been found. Nevertheless, different approaches have been extensively investigated in order to develop calcium phosphate-based composites. Despite a large number of studies on the synthesis of HAp and TCP composites (Viswanath & Ravishankar, 2006; Rao & Kannan, 2002; Nath et al., 2009; Jin. et al., 2010; Cao & Kuboyama, 2010; Hu et al., 2010), no systematic investigations on the preparation of HAp/MgTiO₃–MgO are performed. Therefore, a novel approach to synthesis of HAp/MgTiO₃–MgO nanocomposite has developed by our research group (Fahami et al., 2012). In this procedure, the starting reactant materials are $CaHPO_4$, CaO, titanium dioxide (TiO_2), and elemental magnesium (Mg). Synthesis of HAp/MgTiO₃–MgO composite nanopowders consists of: (i) mechanical activation of powder mixture, and (ii) subsequently thermal treatment at 700 C for 2 h. The obtained mixture was milled in a high energy planetary ball mill for 10 h according to the following reaction.

$$6CaHPO_4 + 4CaO + TiO_2 + 2Mg \rightarrow Ca_{10}(PO_4)_6(OH)_2 + TiO_2 + 2MgO + 2H_2\uparrow \qquad (9)$$

$$MgO + TiO_2 \rightarrow MgTiO_3$$
$$\Delta G_{298K} = -25.410 \text{ kJ}, \quad \Delta H_{298K} = -26.209 \text{ kJ} \tag{10}$$

Figure 13 shows the XRD profiles of the $CaHPO_4$, CaO, TiO_2, and Mg powder mixture after 10 h of milling and after thermal annealing at 700 C for 2 h. As can be seen, the product of mechanochemical process in presence of 20 wt.% (TiO_2, Mg) is $HAp/MgO–TiO_2$ composite. From Figure 13b, it was verified the existence of HAp and geikielite ($MgTiO_3$) phases together with minor MgO phase after the annealing at 700 ºC. This suggests that the thermal treatment at 700 ºC led to the formation of $MgTiO_3$ by the following reaction:

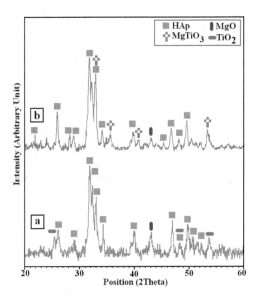

Figure 13. XRD patterns of the samples: (a) $HAp/MgO–TiO_2$ after 10 h of milling, and (b) $HAp/MgTiO_3–MgO$ after 10 h of milling + heat treatment at 700 ºC (Fahami et al., 2012).

Figure 14 shows the SEM micrographs and EDX results of the samples after milling and thermal treatment. As can be seen in Figure 15a, a very fine structure was formed after 10 h of milling. After thermal treatment at 700 ºC (Figure 14c), continuous evolution of the morphological features was appeared. The mean size of the powder particles increased after thermal treatment; however, only a slight change in particle size was observed in heat treated sample at 700 ºC compare to the milled powder. After milling and subsequent thermal treatment at 700 ºC, the products were composed of fine particles with a mean particle size of about 482 and 510 nm, respectively. Figures 14b and d represent the EDX results for HAp-based composites which are synthesized after 10 h of milling and subsequent heat treatment at 700 ºC. EDX data show that the main elements of the calcium phosphate-based composite nanopowders are calcium, phosphorus, oxygen, magnesium, and titanium. The EDX of HAp crystal, present in the HAp/MgO –TiO$_2$ composite exhibit a molar ratio Ca/P = 1.93, whereas in the HAp/MgTiO$_3$–MgO composite the molar ratio of calcium to phosphorus is greater (Ca/P = 2.87). These results suggest that the HAp crystals are closer to the expected value for the molar ratio of calcium to phosphorus ratio for the standard HAp (Ca/P = 1.67) (Cacciotti et al., 2009) and commercial HAp (2.38) (Silva et al., 2002), respectively. It is noteworthy to mention that chemically stable contaminants were not detected due to the excessive adhesion of powders to the vial and balls. Figure 15 demonstrates the TEM images of the HAp/MgTiO$_3$–MgO composite nanopowders produced after 10 h of milling and subsequent annealing at 700 °C for 2h. As can be seen, the agglomerates with mean size of about 322 nm were developed after thermal treatment at 700 °C. In this sample, the cluster-like shape particles were composed of fine spheroidal shape crystals with a mean size of about 55 nm. It should be mentioned that chemical interactions at the contacting surface of crystals resulted in cluster-like shape aggregates which were composed of fine spheroidal shape crystals. This phenomenon is referred to the nature of milling process which originates through repeated welding, fracturing and re-welding of fine powder particles (Suryanarayana, 2001; De Castro & Mitchell, 2002). It is found that MgO-doped HAp/TCP ceramics present high density and significantly enhance mechanical properties without any phase transformation of β-TCP to α-TCP up to 1300 °C (Farzadi et al., 2011). Moreover, a patent (Sul, 2008) reported the biocompatibility and osteoconductivity of magnesium titanate oxide film implant for utilizing in several medical fields such as dentistry, orthopedic, maxillofacial, and plastic surgery. Therefore, the presence of MgO and MgTiO$_3$ phases along with HAp in outputs can be enhanced the biological and mechanical properties of HAp-based bioceramics.

3.2. Mechanochemical synthesize of fluorapatite nanostructures

The inorganic matrix of the bone is based on HAp doped with different quantities of cations, such as Na$^+$, K$^+$ and Mg^{2+}, and anions, such as CO$_3$$^{2-}$, SO$_4$$^{2-}$ and F$^-$. Among them, F$^-$ plays a leading role because of its influence on the physical and biological characteristics of HAp (Nikcevic et al., 2004). In the recent years, fluoridated HAp (FHAp and FAp) has attracted much attention as a promising material to replace HAp in biomedical applications (Kim et al., 2004a; Fathi & Mohammadi Zahrani, 2009). It is found that the incorporation of fluoride ions into the HAp structure considerably increases the resistance of HAp to biodegradation and thermal decomposition (Fathi et al., 2009). In addition, fluoridated hydroxyapatite could provide better protein adsorption (Zeng et al., 1999) and comparable or better cell attachment than HAp (Kim et al., 2004b). This substitution also has positive effects on proliferation, morphology and differentiation of osteoblastic-like cells and promotes the bioactivity (Fathi et al., 2009). For all these reasons, synthesis of FHAp and FAp is of great value and has been widely investigated by multiple techniques, such as precipitation (Chen & Miao, 2005), sol–gel (Cheng et al., 2006), hydrolysis (Kurmaev et al., 2002), hydrothermal (Sun et al., 2012), and mechanochemical methods (Nikcevic et al., 2004).

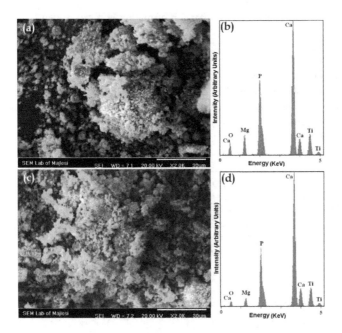

Figure 14. SEM micrographs and EDX results of the samples: (a-b) HAp/MgO–TiO$_2$ after 10 h of milling (c-d) HAp/ MgTiO$_3$–MgO after 10 h of milling + heat treatment at 700 °C for 2 h.

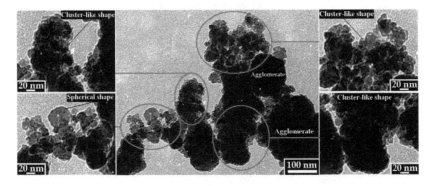

Figure 15. TEM images of the HAp/M$_g$TiO$_3$–MgO composite nanopowders after 10 h of milling + heat treatment at 700 °C for 2 h (Fahami et al., 2012).

Nikcevic et al. (Nikcevic et al., 2004) synthesized nanostructured fluorapatite/fluorhydrox-yapatite and carbonated fluorapatite/fluorhydroxyapatite by mechanochemical process. Powder mixture of Ca(OH)$_2$-P$_2$O$_5$-CaF$_2$ were milled in planetary ball mill. A carbonated fluorhydroxyapatite, FHAp was formed after 5 h of milling and carbonated fluoroapatite was formed after 9 h of milling. Complete transformation of the carbonated form of FAp into the single phase of FAp occurred after 9 h milling and thermally treating. After that, Zhang et al. (Zhang et al., 2005) synthesized FHAp from the starting materials of $CaCO_3$, $CaHPO_4.2H_2O$, and CaF_2 via a mechanochemical–hydrothermal route. The mechanism study revealed that under such mechanochemical–hydrothermal conditions the formation reactions of FHAp were completed in two stages. The starting materials firstly reacted into a poorly crystallized calcium-deficient apatite and the complete incorporation of fluoride ions into apatite occurred in the second stage.

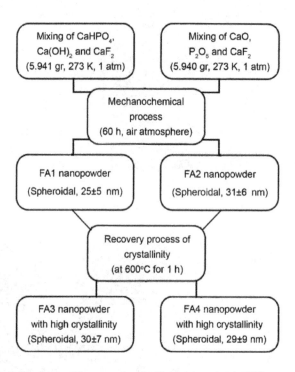

Figure 16. Flow sheet of FAp nanoparticles preparation (Ebrahimi-Kahrizsangi et al., 2011).

3.2.1. Single-crystal fluorapatite nanoparticles

Along with the development of mechanochemical processes, Mohammadi Zahrani & Fathi (Mohammadi Zahrani & Fathi, 2009) evaluated the effect of ball milling parameters on the synthesis of FAp nanopowder; also, the effect of fluoridation on bioresorbability and bioactivity of apatite was studied. Fluoridated hydroxyapatite nanopowders with 100% (FAp) were synthesized via mechanical alloying method. The results showed that the size and number of balls had no significant effect on the synthesizing time and grain size of FAp, while decreasing the rotation speed or ball to powder weight ratio increased synthesizing time and the grain size of FAp. In vitro test indicated that the bioactivity of FAp was less than HAp since the dissolution rate, precipitation amount and the size of precipitated bone-like apatite crystals on the surface of FAp samples was clearly lower than HAp (Nikcevic et al., 2004; Zhang et al., 2005). Recently, Ebrahimi-Kahrizsangi et al. (Ebrahimi-Kahrizsangi et al., 2011) synthesized FAp nanostructures from the starting materials of $CaHPO_4$, $Ca(OH)_2$, CaO, P_2O_5 and CaF_2 via mechanochemical process. The suitability of using the mechano-chemical process to prepare a high crystalline phase of FAp was studied. FAp nanopowders with different structural characteristics synthesized through novel dry mechanochemical processes, which are presented in Figure 16.

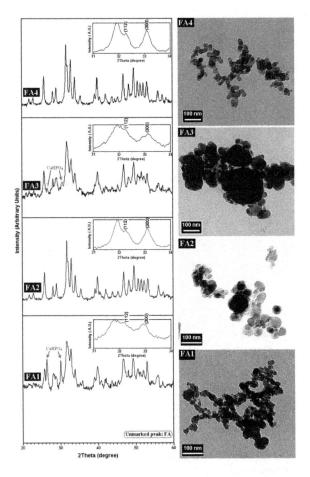

Figure 17. XRD patterns and morphological features of the FAp nanostructures (Ebrahimi-Kahrizsangi et al., 2011).

The purpose of the milling was twofold: first, to activate the following reactions via mechanochemical processes, and second, to produce the nanostructured FAp.

$$6CaHPO_4 + 3Ca(OH)_2 + CaF_2 \rightarrow Ca_{10}(PO_4)_6F_2 + 6H_2O \qquad (11)$$

$$9CaO + 3P_2O_5 + CaF_2 \rightarrow Ca_{10}(PO_4)_6F_2 \qquad (12)$$

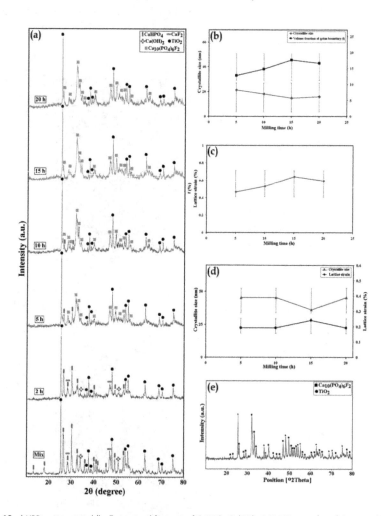

Figure 18. a) XRD patterns, and (b-d) structural features of $CaHPO_4$-$Ca(OH)_2$-CaF_2-TiO_2 powder mixture, mechanically alloyed for 2-20 h. (e) XRD profile of the FAp–TiO_2 nanocomposite after heat treatment at 650 C (Ebrahimi-Kahrizsangi et al., 2011).

Figure 17 illustrates XRD patterns and morphological features of the FAp nanostructures. XRD patterns of the FA1 and FA2 samples showing that the powders synthesized through the two different mechanochemical processes are mostly FAp. Complete agreement with the standard card of fluorapatite (JCPDS #15-0876) was not observed for FA1 due to the presence of additional peaks at $2\theta = 26.59\circ$ and $30.19\circ$. These additional peaks were attributed to $CaHPO_4$ from the starting materials. In FA2, complete agreement with JCPDS #15-0876 was observed, allowing FA2 to be used as the pure FAp phase sample when required. The results from the structural studies indicate that the maximum lattice disturbance in the apatite structure after the mechanochemical process was at the $(0\,0\,2)$ plane. According to TEM images, the FA1 particles are spheroidal with an average diameter of 25 ± 5 nm. However, it should be noted that the particles do not possess high regularity in shape; in other words, their surfaces are not smooth. The particles show a high tendency towards agglomeration. In addition, the sample FA2 possesses a mostly spherical structure with an average diameter of 31 ± 6 nm. The TEM image of the FA3 sample confirms that the particles are spheroidal with an average diameter of 30 ± 7 nm. As shown in Figure 17, the TEM image of sample FA4 shows the particles to be more spherical than the unheated FA2 particles, and the average size of the particles is 29 ± 9 nm. Based on the obtained data, the maximum particle size measured with TEM is below the crystallite size calculated from the line broadening of the X-ray diffraction peak. Thus we concluded that, after 60 h of milling and subsequent thermal treatment at 600 C, this method gives rise to single-crystal FAp with average sizes of 30, 37, 37 and 38 nm for FA1, FA2, FA3 and FA4, respectively. In most reports concerning the synthesis of FAp, the particle shapes are plates (Rameshbabu et al., 2006) or polyhedral (Barinov et al., 2004; Fathi & Mohammadi Zahrani, 2009), but nanoparticles with spheroidal morphology were successfully prepared by current approach. Because the spherical geometry rather than irregular shape is important for achieving osseointegration (Hsu et al., 2007; Nayar et al., 2006), the products synthesized via mechanochemical processes are preferred for medical applications.

3.2.2. Fluorapatite-titania (FAp-TiO₂) nanocomposite

As a fact that the incorporation of bioinert ceramics into calcium phosphate-based materials has demonstrated significant improvement in mechanical properties without substantial compromise in biocompatibility, some attempts have been made to develop FHAp-based composites such as: $FHAp–Al_2O_3$ (Adolfsson et al., 1999), and $FHAp–ZrO_2$ (Kim et al., 2003; Ben Ayed & Bouaziz, 2008) composites. However, only a few studies have been devoted to the use of solid state reaction in order to prepare FAp nanocomposites (Bouslama et al., 2009). Therefore, synthesis of $FAp–TiO_2$ nanocomposite which can present advantages of both TiO_2 and FAp were carried out by Ebrahimi-Kahrizsangi et al. (Ebrahimi-Kahrizsangi et al., 2011). Based on XRD patterns and FT-IR spectroscopy, correlation between the structural features of the nanostructured $FAp–TiO_2$ and the process conditions was investigated. The starting reactant materials are $CaHPO_4$, $Ca(OH)_2$, CaF_2, and TiO_2. In the production of the nanocomposite, a distinct amount of titanium dioxide (20 wt.%) was mixed with CaH-PO$_4$, Ca(OH)$_2$ and CaF$_2$ according to reaction (13), and were milled in planetary ball mill for 2, 5, 10, 15, and 20 h under ambient air atmosphere. The aims of the milling were twofold: the first one was to activate the following reaction via one step mechanochemical process, and the secondly, was to produce the FAp–TiO$_2$ nanocomposite.

$$6CaHPO_4 + 3Ca(OH)_2 + CaF_2 + TiO_2 \rightarrow Ca_{10}(PO_4)_6F_2 + TiO_2 + 6H_2O \qquad (13)$$

Figure 18a shows the XRD patterns of CaHPO$_4$–Ca(OH)$_2$–CaF$_2$–TiO$_2$ powder mixture, mechanically alloyed for 2-20 h. An XRD pattern of the mixture before milling is given in the same figure for comparison. For the powder mixture, milled for 2 h, all the sharp peaks corresponding to Ca(OH)$_2$, CaF$_2$ have diminished, and those corresponding to CaHPO$_4$ have been broadened, indicating that a significant refinement in crystallite and particle sizes of the starting powders had occurred together with a degree of amorphization at the initial stage of mechanical activation. Also, the X-ray pattern of the sample, milled for 2 h, shows the most intense peaks for TiO$_2$. Upon 5 h of mechanical activation, several new broadened peaks especially between 2θ = 31 –34 appear to emerge, corresponding to FAp phase. This suggests that nanocrystalline FAp phase has been formed as a result of mechanical activation. According to XRD profile, the main products of mechanochemical process after 5 h of milling were FAp and TiO$_2$. Also, the two minor peaks observed in XRD patterns correspond to CaHPO$_4$. When the mechanical activation time is extended to 15 h, all the peaks corresponding to CaHPO$_4$ have disappeared and only those belonging to FAp and TiO$_2$ are detectable.

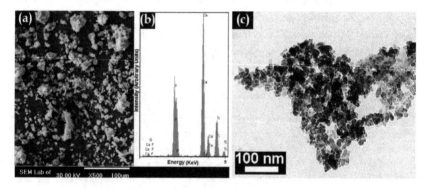

Figure 19. SEM micrograph and TEM image of FAp–TiO$_2$ nanocomposite after 20 h of milling (Ebrahimi-Kahrizsangi et al., 2011).

Figures 18b and c show the variations of the crystallite size, lattice strain, and the volume fraction of grain boundaries of FAp as a function of milling time. Mechanical activation up to 15 h leads to a rapid decrease in the crystallite size to less than 16 nm, and a large increase in the volume fraction of grain boundary to 17.55%. Upon 15 h of mechanical activation, the crystallite size and the volume fraction of grain boundary reach about 16 nm and 16.54%, respectively. The evaluation of the lattice strain indicates that the lattice strain significantly increased with mechanical activation until 15 h, and then decreased slightly with further milling up to 20 h. The determined amounts of the crystallite size and the lattice strain of TiO_2 as reinforcement are presented in Figure 18 d. The crystallite size of TiO_2 remains near-ly constant with milling until 10 h, and then decreases severely with further mechanical acti-vation up to 15 h. After 20 h of milling, the lattice strain decreases which leads to an increase in the crystallite size. According to data presented in Figure 18 with further mechanical acti-vation up to 20 h for both matrix and reinforcement, the lattice strain can be decreased at higher milling intensities because of the enhanced dynamical recrystallisation. Because the calcium phosphate ceramics with higher crystallinity degree has lower activity towards bio-resorption and lower solubility in physiological environment (Sanosh et al., 2009), thermal recovery of crystallinity was performed at 650 C for 2 h. Some increase in the peak intensity of the crystalline FAp phase was observed in the XRD pattern (Figure 18e). The crystalline phase of titanium dioxide (TiO_2) similarly appeared upon annealing. After heat treatment, the breadth of the fundamental diffraction peaks decrease as compared to the results of the milled powder which can be attributed to an increase in crystallite size and a decrease in lattice strain. The determined amounts of the structural features indicated that the crystallite size and lattice strain of FAp reached 43.3 nm and 0.30%, respectively.

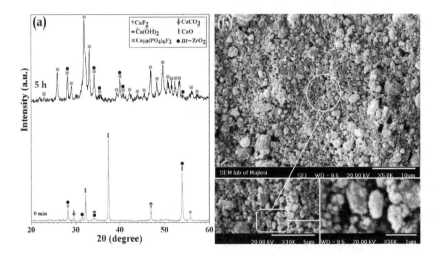

Figure 20. XRD patterns and morphological features of the FAp-ZrO_2 composite nanopowders after 5 h of milling.

The SEM micrograph, EDX result, and TEM image of the mechanosynthesized FAp–20 wt. %TiO_2 nanocomposite after 20 h of mechanical activation time are shown in Figure 19. After 20 h of milling, the mean size of powder particles decreased due to severe mechanical defor-mation introduced into the powder. It also emerged from SEM image, that the fine agglomerates consist of significantly finer agglomerates/particles that cannot be seen individually in micrographs because of their exceptionally small sizes. Therefore, the size and morphology of fine powders was determined by using transmission electron microscopy. The results of measurements of elemental composition by EDX confirm that very homogeneous distribution of components is formed during one step mechanochemical process particularly after 20 h of milling. The results of EDX analysis also reveal that no chemically stable contaminants are detected due to the excessive adhesion of powders to the milling media. From TEM images it is clearly seen that the synthesized powder after 20 h of milling has an appropriate homogeneity. Also, the particles of products are in average size of about 15 nm after 20 h of milling, respectively. Due to high surface energy, the nanostructured materials can improve the sinterability and, thus, improve mechanical properties (Suryanarayana, 2001). Of course, sintering behavior not only depends on particle size, but also on particle size distribution and morphology of the powder particles. Large particle size along with hard agglomerates shows lower densification in calcium phosphate ceramics. On the other hand, difference in shrinkage between the agglomerates is also responsible to produce small cracks in the sintered calcium phosphate-based ceramic (Banerjee et al., 2007). Therefore, preparation of agglomerate free or soft agglomerated nanostructured FAp–20 wt.%TiO_2 can be an important parameter to achieve good mechanical properties for dense nanostructure. The results of SEM and TEM images suggest that the formation of FAp–20 wt.%TiO_2 nanocomposite after 20 h of milling does not accompany hard agglomerates and, therefore, the synthesized powder can present appropriate mechanical properties.

3.2.3. Fluorapatite-Zirconia (FAp-ZrO₂) nanocomposite

3.2.3. Fluorapatite-Zirconia (FAp-ZrO_2) nanocomposite

Among the reinforcement materials of ceramic-based bionanocomposites, ZrO_2 as a bioinert reinforcement has been studied extensively because of its relatively higher mechanical strength and toughness (Rao & Kannan, 2002; Ben Ayed & Bouaziz, 2008; Evis, 2007). However, the addition of ZrO_2 results in lowering the decomposition temperature of microcrystalline HAp- and FHAp-based composites below the sintering temperature which causes an adverse influence on the mechanical properties (Kim et al., 2003). These phenomena are related to structural features of composite that are affected by the synthesis process. Generally, in order to prepare HAp/FHAp–ZrO_2 composites, calcium phosphate source chemicals and ZrO_2 powders are mixed, cold pressed and then sintered at high temperatures (Rao & Kannan, 2002). Under these conditions, the resulting product could have microcrystalline structure. Since the nanocrystalline structure compared to the microcrystalline structure is more important to achieve high thermal stability and mechanical properties, the FAp–ZrO_2 composites with nanostructural characteristics are preferred for medical applications. In fact, the composite nanopowders can improve the sinterability as well as mechanical properties due to high surface energy. For all these reasons, our research group was considered to synthesis of FAp–ZrO_2 nanocomposite with appropriate structural features via one step mechanochemical process. For this aim, commercially available calcium oxide (CaO), phosphorous pentoxide (P_2O_5), calcium fluoride (CaF_2), and monoclinic zirconia (m–ZrO_2) were used as starting reagents. The mechanochemical synthesis was performed in a planetary ball mill without using any process control agent (PCA).

Figure 20 shows the XRD patterns and SEM images of FAp-ZrO_2 composite nanopowders after mechanical activation for 5 h. An XRD pattern of the mixture before milling is also given in the same figure for comparison. As can be seen in this figure, only sharp characteristic peaks of CaO and CaF_2 could be detected at the initial mechanical activation of starting powder mixture. Phosphoric acid was formed immediately upon addition of P_2O_5 to the reaction mixture due to very high hydrophilic of P_2O_5. Therefore, characteristic peaks of P_2O_5 could not be observed. On the other hand, the CaO is not stable and will spontaneously react with H_2O and CO_2 from the air that leading to formation of $Ca(OH)_2$ and $CaCO_3$ in powder mixture. According to the XRD patterns, it can be seen that after 5 h of milling, all the peaks corresponding to CaO have vanished and only those belonging to FAp and m–ZrO_2 were visible. Figure 20b shows the morphological characteristics of the FAp-ZrO_2 composite nanopowders after 5 h of milling at different magnifications. According to this figure, a homogeneous microstructure was obtained after 5 h of milling which is important for the improvement in mechanical properties.

4. Conclusion and future directions

Hydroxyapatite- and fluorapatite-based nanocomposite powders have been developed and demonstrate huge potential for a variety of biomedical applications such as controlled drug release, bone cements, tooth paste additive, and dental implants. Different types of calcium phosphate-based nanocomposites can be synthesized by various approaches for instance wet chemical methods, hydrothermal processes, solid–state reaction, sol–gel method, and mechanochemical processes. Compared to various synthesis processes, mechanosynthesis method present a number of advantages such as high efficiency and enabling to synthesis a wide range of novel advanced materials (nanocomposites). It has been proved that the structural, mechanical, and biological properties of bioceramics can be significantly improved by using calcium phosphate-based nanocomposites as the advanced materials. For this reason, many attempts have been made to improve the mechanical properties as well as structural features of bioceramics through the incorporation of ceramic second phases. Mechanochemical synthesis of hydroxyapatite- and fluorapatite-based nanocomposite powders are of great interest and should be further explored. Although mechanochemical process have demonstrated their great potential for synthesis of hydroxyapatite- and fluorapatite-based nanocomposite powders, several challenges still remain. Mechanosynthesis of nanocomposites with precisely controlling their chemical composition, phases, biological characteristics, mechanical properties, and interfacial features is still a challenging task. Depending on the preparation circumstances (milling atmosphere, milling time, milling temperature, and type of milling media) and chemical composition of initial materials, the properties and performance of the nanocomposites can vary significantly; therefore, the ability to reproduce calcium phosphate-based nanocomposites with unique characteristics is very important for their wide use as biomaterials. With the increased interests and intensive research and development in the field of mechanochemistry, it is expected that, mechanosynthesized nanocomposites will have a promising future and will make a significant influence on the advanced materials industry.

Acknowledgements

The authors are grateful to research affairs of Islamic Azad University, Najafabad Branch and Iranian Nanotechnology Initiative Council (INIC) for supporting this research.

Author details

Bahman Nasiri–Tabrizi[1], Abbas Fahami[1], Reza Ebrahimi–Kahrizsangi[1] and Farzad Ebrahimi[2]

1 Materials Engineering Department, Najafabad Branch, Islamic Azad University, Najafabad, Isfahan, Iran

2 Department of Mechanical Engineering, Faculty of Engineering, Imam Khomeini International University, Qazvin, Iran

References

[1] Adamopoulos, O., & Papadopoulos, T. (2007). Nanostructured bioceramics for maxillofacial applications. *Journal of Materials Science: Materials in Medicine*, 18, 1587-1597.

[2] Adolfsson, E., Nygren, N., & Hermansson, L. (1999). Decomposition mechanisms in aluminum oxide-apatite systems. *Journal of the American Ceramic Society*, 82, 2909-2912.

[3] Avvakumov, E., Senna, M., & Kosova, N. (2002). *Soft mechanochemical synthesis: A basis for new chemical technologies,*, Kluwer Academic Publishers.

[4] Balamurugan, A., Kannan, S., & Rajeswari, S. (2002). Bioactive sol-gel hydroxyapatite surface for biomedical application-in vitro study. *Trends in Biomaterials & Artificial Organs*, 16, 18-20.

[5] Balaz, P. (2008). *Mechanochemistry in Nanoscience and Minerals Engineering*, Springer.

[6] Banerjee, A., Bandyopadhyay, A., & Bose, S. (2007). Hydroxyapatite nanopowders: synthesis, densification and cell materials interaction. *Materials Science and Engineering C*, 27, 729-735.

[7] Barinov, S. M., Shvorneva, L. I., Ferro, D., Fadeeva, I. V., & Tumanov, S. V. (2004). Solid solution formation at the sintering of hydroxyapatite-fluorapatite ceramics. *Science and Technology of Advanced Materials*, 5, 537-541.

[8] Ben, Ayed. F., & Bouaziz, J. (2008). Sintering of tricalcium phosphate-fluorapatite composites with zirconia. *Journal of the European Ceramic Society*, 28, 1995-2002.

[9] Bose, S., Xue, W., Banerjee, A., & Bandyopadhyay, A. (2009). Spherical and anisotropic hydroxyapatite nanocrystals Challa S. S. R. Kumar ed. *Nanomaterials for the Life Sciences 2, Nanostructured Oxides*, WILEY-VCH Verlag GmbH & Co, 407-447.

[10] Bouslama, N., Ben, Ayed. F., & Bouaziz, J. (2009). Sintering and mechanical properties of tricalcium phosphate-fluorapatite composites. *Ceramics International*, 35, 1909-1917.

[11] Cacciotti, I., Bianco, A., Lombardi, M., & Montanaro, L. (2009). Mg-substituted hydroxyapatite nanopowders: Synthesis, thermal stability and sintering behaviour. *Journal of the European Ceramic Society*, 29, 2969-2978.

[12] Cao, H., & Kuboyama, N. (2010). A biodegradable porous composite scaffold of PGA/β-TCP for bone tissue engineering. *Bone*, 46, 386-395.

[13] Chen-W, Ch., Riman, R. E., Ten, Huisen. K. S., & Brown, K. (2004). Mechanochemical-hydrothermal synthesis of hydroxyapatite from nonionic surfactant emulsion precursors. *Journal of Crystal Growth*, 270, 615-623.

[14] Cheng, K., Zhang, S., & Weng, W. (2006). Sol-gel preparation of fluoridated hydroxyapatitein Ca(3 $_2$-PO(OH)$_{3-x}$(OEt)$_x$-HPF$_6$ system. *Journal of Sol-Gel Science and Technology*, 38, 13-17.

[15] Chen, Y., & Miao, X. (2005). Thermal and chemical stability of fluorohydroxyapatite ceramics with different fluorine contents. *Biomaterials*, 26, 1205-1210.

[16] Choi-Y, W., Kim-E, H., Oh, S. Y., & Koh, Y. H. (2010). Synthesis of poly(ϵ-caprolactone)/hydroxyapatite nanocomposites using in-situ co-precipitation. *Materials Science and Engineering C*, 30, 777-780.

[17] De Castro, C. L., & Mitchell, B. S. (2002). Synthesis functionalization and surface treatment of nanoparticles. *Nanoparticles from mechanical attrition*, Stevenson Ranch, CA: American Scientific Publishers, 1-14.

[18] Ducheyne, P., & Qiu, Q. (1999). Bioactive ceramics: the effect of surface reactivity on bone formation and bone cell function. *Biomaterials*, 20, 2287-2303.

[19] Ebrahimi-Kahrizsangi, R., Nasiri-Tabrizi, B., & Chami, A. (2010). Synthesis and characterization of fluorapatite-titania (FAp-TiO$_2$) nanocomposite via mechanochemical process. *Solid State Sciences*, 12, 1645-1651.

[20] Ebrahimi-Kahrizsangi, R., Nasiri-Tabrizi, B., & Chami, A. (2011). Characterization of single-crystal fluorapatite nanoparticles synthesized via mechanochemical method. *Particuology*, 9, 537-544.

[21] El Briak-Ben, Abdeslam. H., Ginebra, M. P., Vert, M., & Boudeville, P. (2008). Wet or dry mechanochemical synthesis of calcium phosphates? Influence of the water content on DCPD-CaO reaction kinetics. *Acta Biomaterialia*, 4, 378-386.

[22] Enayati-Jazi, M., Solati-Hashjin, M., Nemati, A., & Bakhshi, F. (2012). Synthesis and characterization of hydroxyapatite/titania nanocomposites using in situ precipitation technique. *Superlattices and Microstructures*, 51, 877-885.

[23] Evis, Z. (2007). Reactions in hydroxylapatite-zirconia composites. *Ceramics International*, 33, 987-991.

[24] Fahami, A., Ebrahimi-Kahrizsangi, R., & Nasiri-Tabrizi, B. (2011). Mechanochemical synthesis of hydroxyapatite/titanium nanocomposite. *Solid State Sciences*, 13, 135-141.

[25] Fahami, A., Nasiri-Tabrizi, B., & Ebrahimi-Kahrizsangi, R. (2012). Synthesis of calcium phosphate-based composite nanopowders by mechanochemical process and subsequent thermal treatment. *Ceramics International*.

[26] Farzadi, A., Solati-Hashjin, M., Bakhshi, F., & Aminian, A. (2011). Synthesis and characterization of hydroxyapatite/β-tricalcium phosphate nanocomposites using microwave irradiation. *Ceramics International*, 37, 65-71.

[27] Fathi, M. H., & Mohammadi Zahrani, E. (2009). Fabrication and characterization of fluoridated hydroxyapatite nanopowders via mechanical alloying. *Journal of Alloys and Compounds*, 475, 408-414.

[28] Fathi, M. H., Mohammadi Zahrani, E., & Zomorodian, A. (2009). Novel fluorapatite/niobium composite coating for metallic human body implants. *Materials Letters*, 63, 1195-1198.

[29] Fini, M., Savarino, L., Nicoli, Aldini. N., Martini, L., Giavaresi, G., Rizzi, G., Martini, D., Ruggeri, A., Giunti, A., & Giardino, R. (2003). Biomechanical and histomorphometric investigations on two morphologically differing titanium surfaces with and without fluorohydroxyapatite coating: An experimental study in sheep tibiae. *Biomaterials*, 24, 3183-3192.

[30] Gergely, G., Wéber, F., Lukács, I., Tóth, A. L., Horváth, Z. E., Mihály, J., & Balázsi, C. (2010). Preparation and characterization of hydroxyapatite from eggshell. *Ceramics International*, 36, 803-806.

[31] Gonzalez, G., Sagarzazu, A., & Villalba, R. (2006). Mechanochemical transformation of mixtures of $Ca(OH)_2$ and $(NH_4)_2HPO_4$ or 2O5. *Materials Research Bulletin*, 41, 1902-1916.

[32] Gu, M., Huang, C., Xiao, S., & Liu, H. (2008). Improvements in mechanical properties of TiB_2 ceramics tool materials by dispersion of Al_2O_3 particles. *Materials Science and Engineering A*, 486, 167-170.

[33] Gu, Y. W., Loh, N. H., Khor, K. A., Tor, S. B., & Cheang, P. (2002). Spark plasma sintering of hydroxyapatite powders. *Biomaterials*, 23, 37-43.

[34] Honarmandi, P., Honarmandi, P., Shokuhfar, A., Nasiri-Tabrizi, B., & Ebrahimi-Kahrizsangi, R. (2010). Milling media effects on synthesis, morphology and structural

characteristics of single crystal hydroxyapatite nanoparticles. *Advances in Applied Ceramics*, 109, 117-122.

[35] Hsu, Y. H., Turner, I. G., & Miles, A. W. (2007). Fabrication and mechanical testing of porous calcium phosphate bioceramic granules. *Journal of Materials Science: Materials in Medicine*, 18, 1931-1937.

[36] Hu, H., Liu, X., & Ding, Ch. (2010). Preparation and in vitro evaluation of nanostructured TiO$_2$/TCP composite coating by plasma electrolytic oxidation. *Journal of Alloys and Compounds*, 498, 172-178.

[37] Jallot, E., Nedelec, J. M., Grimault, A. S., Chassot, E., Grandjean-Laquerriere, A., Laquerriere, P., & Laurent-Maquin, D. (2005). STEM and EDXS characterisation of physico-chemical reactions at the periphery of sol-gel derived Zn-substituted hydroxyapatites during interactions with biological fluids. *Colloids and Surfaces B: Biointerfaces*, 42, 205-210.

[38] Jin, H. H., Min, S. H., Song, Y. K., Park, H., Ch, , & Yoon, S. Y. (2010). Degradation behavior of poly(lactide-co-glycolide)/β-TCP composites prepared using microwave energy. *Polymer Degradation and Stability*, 95, 1856-1861.

[39] Kalita, S. J., Bhardwaj, A., & Bhatt, H. A. (2007). Nanocrystalline calcium phosphate ceramics in biomedical engineering. *Materials Science and Engineering C*, 27, 441-449.

[40] Kano, J., Zhang, Q., Saito, F., Baron, M., & Nzihou, A. (2006). Synthesis of hydroxyapatite with the mechanochemical treatment products of PVC and CaO Trans I Chem E, Part B. *Process Safety and Environmental Protection*, 84(B4), 309-312.

[41] Khaghani-Dehaghani, M. A., Ebrahimi-Kahrizsangi, R., Setoudeh, N., & Nasiri-Tabrizi, B. (2011). Mechanochemical synthesis of Al$_2$O$_3$-TiB$_2$ nanocomposite powder from Al-TiO$_2$-H$_3$BO$_3$ mixture. *Int. Journal of Refractory Metals and Hard Materials*, 29, 244-249.

[42] Kim, H. W., Kong, Y. M., Bae, Ch. J., Noh, Y. J., & Kim, H. E. (2004b). Sol-gel derived fluor-hydroxyapatite biocoatings on zirconia substrate. *Biomaterials*, 25, 2919-2926.

[43] Kim, H. W., Kong, Y. M., Koh, Y. H., & Kim, H. E. (2003). Pressureless sintering and mechanical and biological properties of fluor-hydroxyapatite composites with zirconia. *Journal of the American Ceramic Society*, 86, 2019-2026.

[44] Kim, H. W., Li, L. H., Koh, Y. H., Knowles, J. C., & Kim, H. E. (2004a). Sol-Gel preparation and properties of fluoride-substituted hydroxyapatite powders. *Journal of the American Ceramic Society*, 87, 1939-1944.

[45] Kivrak, N., & Tas, A. C. (1998). Synthesis of calcium hydroxyapatite-tricalcium phosphate composite bioceramic powders and their sintering behavior. *Journal of the American Ceramic Society*, 81, 2245-2252.

[46] Komlev, V. S., Barinov, S. M., Orlovskii, V. P., & Kurdyumov, S. G. (2001). Porous Ceramic Granules of Hydroxyapatite. *Refractories and Industrial Ceramics*, 42, 242-244.

[47] Kong, Y. M., Kim, S., & Kim, H. E. (1999). Reinforcement of hydroxyapatite biocer-
      amic by addition of ZrO_2 coated with Al_2O_3 . *Journal of the American Ceramic Society*,
      82, 2963-2968.

[48] Kurmaev, E. Z., Matsuya, S., Shin, S., Watanabe, M., Eguchi, R., Ishiwata, Y., Takeu-
      chi, T., & Iwami, M. (2002). Observation of fluorapatite formation under hydrolysis
      of tetracalcium phosphate in the presence of KF by means of soft X-ray emission and
      absorption spectroscopy. *Journal of Materials Science: Materials in Medicine*, 13, 33-36.

[49] Landi, E., Tampieri, A., Celotti, G., & Sprio, S. (2000). Densification behavior and
      mechanisms of synthetic hydroxyapatites. *Journal of the European Ceramic Society*, 20,
      2377-2387.

[50] Lee-H, H., Shin, U. S., Won-E, J., & Kim-W, H. (2011). Preparation of hydroxyapatite-
      carbon nanotube composite nanopowders. *Materials Letters*, 65, 208-211.

[51] Liu, F., Wang, F., Shimizu, T., Igarashi, K., & Zhao, L. (2006). Hydroxyapatite forma-
      tion on oxide films containing Ca and P by hydrothermal treatment. *Ceramics Interna-
      tional*, 32, 527-531.

[52] Marchi, J., Greil, P., Bressiani, J. C., Bressiani, A., & Müller, F. (2009). Influence of
      synthesis conditions on the characteristics of biphasic calcium phosphate powders.
      International Journal of Applied Ceramic Technology, 6, 60-71.

[53] Mayera, I., & Featherstone, J. D. B. (2000). Dissolution studies of Zn-containing carbo-
      nated hydroxyapatites. *Journal of Crystal Growth.*, 219, 98-101.

[54] Mobasherpour, I., Solati-Hashjin, M., & Kazemzadeh, A. (2007). Synthesis of nano-
      crystalline hydroxyapatite by using precipitation method. *Journal of Alloys and Com-
      pounds*, 430, 330-333.

[55] Mochales, C., El Briak-Ben, Abdeslam. H., Ginebra, M. P., Terol, A., Planell, J. A., &
      Boudeville, Ph. (2004). Dry mechanochemical synthesis of hydroxyapatites from
      DCPD and CaO: influence of instrumental parameters on the reaction kinetics. *Bio-
      materials*, 25, 1151-1158.

[56] Mishra, S. K., Das, S. K., & Pathak, L. C. (2006). Sintering behavior of self-propagat-
      ing high temperature synthesized ZrB_2-Al_2O_3 composite powder. *Materials Science
      and Engineering A*, 426, 229-34.

[57] Mohammadi, Zahrani. E., & Fathi, M. H. (2009). The effect of high-energy ball mill-
      ing parameters on the preparation and characterization of fluorapatite nanocrystal-
      line powder. *Ceramics International*, 35, 2311-2323.

[58] Nakano, T., Tokumura, A., & Umakoshi, Y. (2002). Variation in crystallinity of hy
      droxyapatite and the related calcium phosphates by mechanical grinding and subse-
      quent heat treatment. *Metallurgical and Materials Transactions A*, 33, 521-528.

[59] Nasiri-Tabrizi, B., Honarmandi, P., Ebrahimi-Kahrizsangi, R., & Honarmandi, P. (2009). Synthesis of nanosize single-crystal hydroxyapatite via mechanochemical method. *Materials Letters*, 63, 543-546.

[60] Nath, S., Tripathi, R., & Basu, B. (2009). Understanding phase stability, microstructure development and biocompatibility in calcium phosphate-titania composites, synthesized from hydroxyapatite and titanium powder mixture. *Materials Science and Engineering C*, 29, 97-107.

[61] Nayar, S., Sinha, M. K., Basu, D., & Sinha, A. (2006). Synthesis and sintering of biomimetic hydroxyapatite nanoparticles for biomedical applications. *Journal of Materials Science: Materials in Medicine*, 17, 1063-1068.

[62] Nikcevic, I., Jokanovic, V., Mitric, M., Nedic, Z., Makovec, D., & Uskokovic, D. (2004). Mechanochemical synthesis of nanostructured fluorapatite/fluorhydroxyapatite and carbonated fluorapatite/fluorhydroxyapatite. *Journal of Solid State Chemistry*, 177, 2565-2574.

[63] Otsuka, M., Matsuda, Y., Hsu, J., Fox, J. L., & Higuchi, W. I. (1994). Mechanochemical synthesis of bioactive material: effect of environmental conditions on the phase transformation of calcium phosphates during grinding. *Bio-Medical Materials and Engineering*, 4, 357-362.

[64] Palmer, C. A., & Anderson, J. J. B. (2001). Position of the American dietetic association: The impact of fluoride on health. *Journal of the American Dietetic Association*, 101, 126-132.

[65] Pan, H., Tao, J., Yu, X., Fu, L., Zhang, J., Zeng, X., Xu, G., & Tang, R. (2008). Anisotropic demineralization and oriented assembly of hydroxyapatite crystals in enamel: smart structures of biominerals. *Journal of Physical Chemistry B*, 112, 7162-7165.

[66] Pushpakanth, S., Srinivasan, B., Sreedhar, B., & Sastry, T. P. (2008). An in situ approach to prepare nanorods of titania hydroxyapatite (TiO_2-HAp) nanocomposite by microwave hydrothermal technique. *Materials Chemistry and Physics*, 107, 492-498.

[67] Rajkumar, M., Meenakshisundaram, N., & Rajendran, V. (2011). Development of nanocomposites based on hydroxyapatite/sodium alginate: Synthesis and characterisation. *Materials Characterization*, 62, 469-479.

[68] Rameshbabu, N., Sampath, Kumar. T. S., & Prasad, Rao. K. (2006). Synthesis of nanocrystalline fluorinated hydroxyapatite by microwave processing and its in vitro dissolution study. *Bulletin of Material Science*, 29, 611-615.

[69] Ramesh, S., Tan, C. Y., Tolouei, R., Amiriyan, M., Purbolaksono, J., Sopyan, I., & Teng, W. D. (2012). Sintering behavior of hydroxyapatite prepared from different routes. *Materials and Design*, 34, 148-154.

[70] Rao, R. R., & Kannan, T. S. (2002). Synthesis and sintering of hydroxyapatite-zirconia composites. *Materials Science and Engineering C*, 20, 187-193.

[71] Ren, F., Leng, Y., Xin, R., & Ge, X. (2010). Synthesis, characterization and ab initio simulation of magnesium-substituted hydroxyapatite. *Acta Biomaterialia*, 6, 2787-2796.

[72] Rhee, S. H. (2002). Synthesis of hydroxyapatite via mechanochemical treatment. *Biomaterials*, 23, 1147-1152.

[73] Sanosh, K. P., Chu, M. C., Balakrishnan, A., Lee, Y. J., Kim, T. N., & Cho, S. J. (2009). Synthesis of nano hydroxyapatite powder that simulate teeth particle morphology and composition. *Current Applied Physics*, 9, 1459-1462.

[74] Schneider, O. D., Stepuk, A., Mohn, Dirk., Luechinger, Norman. A., Feldman, K., & Stark, W. J. (2010). Light-curable polymer/calcium phosphate nanocomposite glue for bone defect treatment. *Acta Biomaterialia*, 6, 2704-2710.

[75] Silva, C. C., Graca, M. P. F., Valente, M. A., & Sombra, A. S. B. (2007). Crystallite size study of nanocrystalline hydroxyapatite and ceramic system with titanium oxide obtained by dry ball milling. *Journal of Materials Science*, 42, 3851-3855.

[76] Silva, C. C., Pinheiro, A. G., Figueiro, S. D., Goes, J. C., Sasaki, J. M., Miranda, M. A. R., & Sombra, A. S. B. (2002). Piezoelectric properties of collagen-nanocrystalline hydroxyapatite composites. *Journal of Materials Science*, 37, 2061-2070.

[77] Silva, C. C., Pinheiro, A. G., Miranda, M. A. R., Goes, J. C., & Sombra, A. S. B. (2003). Structural properties of hydroxyapatite obtained by mechanosynthesis. *Solid State Sciences*, 5, 553-558.

[78] Silva, C. C., Valente, M. A., Graça, M. P. F., & Sombra, A. S. B. (2004). Preparation and optical characterization of hydroxyapatite and ceramic systems with titanium and zirconium formed by dry high-energy mechanical alloying. *Solid State Sciences*, 6, 1365-1374.

[79] Suchanek, W. L., Byrappa, K., Shuk, P., Riman, R. E., Janas, V. F., & Ten, Huisen. K. S. (2004). Preparation of magnesium-substituted hydroxyapatite powders by the mechanochemical-hydrothermal method. *Biomaterials*, 25, 4647-4657.

[80] Sul, Y. T. (2008). *Osseoinductive magnesium-titanate implant and method of manufacturing the same*, United State Patent Patent No.: US 7,452,566 B2.

[81] Sun, Y., Yang, H., & Tao, D. (2012). Preparation and characterization of Eu^{3+}-doped fluorapatite nanoparticles by a hydrothermal method. *Ceramics International*.

[82] Suryanarayana, C. (2001). Mechanical alloying and milling. *Progress in Materials Science*, 46, 1-184.

[83] Takacs, L. (2002). Self-sustaining reactions induced by ball milling. *Progress in Materials Science*, 47, 355-414.

[84] Takacs, L., Balaz, P., & Torosyan, A. R. (2006). Ball milling-induced reduction of MoS_2 with Al. *Journal of Materials Science*, 41, 7033-7039.

[85] Tian, T., Jiang, D., Zhang, J., & Lin, Q. (2008). Synthesis of Si-substituted hydroxyapatite by a wet mechanochemical method. *Materials Science and Engineering C*, 28, 57-63.

[86] Toriyama, M., Lavaglioli, F., , A., Krajewski, A., Celotti, G., & Piancastelli, A. (1996).
Synthesis of hydroxyapatite-based powders by mechano-chemical method and their
sintering. *Journal of the European Ceramic Society*, 16, 429-436.

[87] Viswanath, B., & Ravishankar, N. (2006). Interfacial reactions in hydroxyapatite/
alumina nanocomposites. *Scripta Materialia*, 55, 863-866.

[88] Wu-C, S., Hsu-C, H., Wu-N, Y., & Ho-F, W. (2011). Hydroxyapatite synthesized from
oyster shell powders by ball milling and heat treatment. *Materials Characterization*, 62,
1180-1187.

[89] Xia, Z. P., Shen, Y. Q., Shen, J. J., & Li, Z. Q. (2008). Mechanosynthesis of molybdenum carbides by ball milling at room temperature. *Journal of Alloys and Compounds*,
453, 185-90.

[90] Yeong, K. C. B., Wang, J., & Ng, S. C. (2001). Mechanochemical synthesis of nanocrystalline hydroxyapatite from CaO and $CaHPO_4$. *Biomaterials*, 22, 2705-2712.

[91] Zeng, H., Chittur, K. K., & Lacefield, W. R. (1999). Analysis of bovine serum albumin
adsorption on calcium phosphate and titanium surfaces. *Biomaterials*, 20, 377-384.

[92] Zhang, H., Zhu, Q., & Xie, Z. (2005). Mechanochemical-hydrothermal synthesis and
characterization of fluoridated hydroxyapatite. *Materials Research Bulletin*, 40,
1326-1334.

[93] Zhou, H., & Lee, J. (2011). Nanoscale hydroxyapatite particles for bone tissue engineering. *Acta Biomaterialia*, 7, 2769-2781.

[94] Hench, Larry L. (1998). Bioceramics. *Journal of the American Ceramic Society*, 81(7),
1705-1728.

Chapter 2

Conducting Polymer Nanocomposites for Anticorrosive and Antistatic Applications

Hema Bhandari, S. Anoop Kumar and S. K. Dhawan

Additional information is available at the end of the chapter

1. Introduction

Intrinsically conducting polymers (ICPs) have been considered for use in various applications. One of the most important applications of these materials which are attracting considerable attention in the most recent times is in corrosion protection of oxidizable metals [1]. The effective use of conducting polymers for corrosion protection of metals can be carried out by different methods; like formulation of polymers with paints, by electro-deposition of conducting polymers onto metal surface and by direct addition of polymers in the corrosive solution as corrosion inhibitors. Coatings on the surface of metals by polymeric materials have been widely used in industries for the protection of these materials against corrosion [2-13]. Some specific conducting polymers like polyaniline and its derivatives, have been found to display interesting corrosion protection properties. In the past decade, the use of polyaniline as anticorrosion coatings had been explored as the potential candidates to replace the chromium-containing materials, which have adverse health and environmental concerns [14-17]. A polymer behaves as a barrier when it exists in the electronically and ionically insulating state. An important feature of the polymer coating in its conductive state is the ability to store large quantity of charge at the interface formed with a passive layer on a metal. This charge can be effectively used to oxidize base metal to form a passive layer. Thus, the conducting polymer film was also capable of maintaining a stationary potential of the protected metal in the passive range [18]. Application of conducting polymers like polyaniline to corrosion protection of metals is, however, subject to some limitations. First, charge stored in the polymer layer (used to oxidize base metal and to produce passive layer) can be irreversibly consumed during the system's redox reactions. Consequently, protective properties of the polymer coating may be lost with time. Also, porosity and anion exchange properties of conducting polymers could be disadvantageous, particularly when it comes to pitting corrosion caused by small aggressive anions (*e.g.*, chlorides) [19]. An interesting alternative is to consider conducting polymer based composite systems. Composite materials play an important role due to their light weight and improved corrosion resistance. These materials usually comprise of a polymer matrix in which fibres and/or small filler particles are thoroughly dispersed. Silicon dioxide particles, for example, comprise one of the common fillers in composite materials such as plastics and films. Conducting polyaniline/inorganic nanocomposites have also attracted more and more attention. A number of different metals and metal oxide particles have so far been encapsulated into the shell of conducting PANI to produce a host of composites materials. These composite materials have shown better mechanical, physical and chemical properties, due to combining the qualities of conducting PANI and inorganic particles [20-22]. Among various inorganic particles, SiO_2 nanoparticles have attractive attention due to their excellent reinforcing properties for polymer materials [23]. However, SiO_2 is an insulator and a lot of works have been done to expand the applications of insulator SiO_2 as fillers and improve the processability of polyaniline [24-28]. Corrosion protective coatings on the mild steel surface by electrochemical deposition of conducting polymers have been extensively studied [29-33]. These types of coatings lack long lasting ability of metals in a corrosive medium. In order to improve the adherence ability and efficiency of conducting polymer coating on the metal surface, the use of liquids paints have also been performed by many researchers [34-38].

Present chapter is based on the preparation of conducting polymer nanocomposites coating onto the mild steel surface by using powder coating techniques. Powder coatings are often used as an alternation to liquids paints finishing or traditional liquid finishing. The key benefits of powder coating techniques are cost effective, environmentally friendly, excellence of finish and performance. This shows single coat finishes with no primer or any other solvent required and high film thickness can be achieved with single coat. Preparation of PANI/SiO_2 nanocomposites was carried out using in-situ polymerization and evaluation of corrosion protection effect for the polyaniline/SiO_2 nanocomposite materials on the mild steel surface. Corrosion protection performance of these composites was compared with that of the polyaniline by performing a series of electrochemical measurements of corrosion potential, polarization resistance, and corrosion current in 1.0 M HCl solution.

In order to improve anticorrosion performance of mid steel in 3.5 % NaCl aqueous medium, preparation of highly hydrophobic polyaniline-SiO_2 nano-composite coating have also been developed. There are several ways to develop hydrophobic surfaces i.e. by electrodeposition method [39-41], solvent casting of polymers [42], layer-by-layer deposition [43-45], chemical vapor deposition [46, 47], dip-coating and/or self-assembly [48-51] and chemical grafting [52-55]. Most of those methods cannot be easily developed and they need very strict conditions of preparation, and low adhesion coatings are often obtained. However, development of highly hydrophobic conducting polymer nanocomposite coatings via powder coating method was found to be highly adhesive, long lasting and more convenient [56, 57].

2. Corrosion study

The corrosion inhibition performance study was carried out at room temperature in aqueous solution of 1.0 M HCl/ 3.5 % NaCl by using Tafel extrapolation and chrono-amperometry methods. Experiments were carried in a conventional three electrode cell assembly using Autolab Potentiostat/ Galvanostat, PGSTAT100 (Nova Software). In three electrode cell assembly, pure iron of dimension 1 cm x 1 cm is taken as working electrode embedded in araldite epoxy, Pt as counter electrode and saturated calomel electrode (SCE) as reference electrode. The cleaning of the working iron electrode was carried out by 1/0, 2/0, 3/0 and 4/0 grade emery papers. The electrodes were then thoroughly cleaned with acetone and tri-chloroethylene to remove any impurities on the surface.

2.1. Tafel extrapolation method

Tafel extrapolation method involves the measurement of over potentials for various current densities. Figure 1 shows a potential vs. log absolute current plot for an applied potential scan. The linear Tafel segments to the anodic and cathodic curves (-0.2 to + 0.2 V versus corrosion potential) were extrapolated to corrosion potential to obtain the corrosion current densities. The slope gives the Tafel slopes (b_a and b_c) and the intercept corresponds to i_{corr}. The corrosion current density [i_{corr} (A/cm^2)] was calculated with the Stern-Geary equation [58] (Eq.1);

$$i_{corr} = \frac{b_a.b_c}{2.3(R_p)(b_a + b_c)} \tag{1}$$

Corrosion rate (C.R) in mm/year can also be calculated by using following relationship [59] (Eq.2);

$$C.R = 3.268 \times 10^3 \frac{i_{corr}}{\rho} \frac{MW}{z} \tag{2}$$

where MW is the molecular weight of the specimen (g/mole), ρ is density of the specimen (g/m^3) and z is the number of electrons transferred in corrosion reactions.

The corrosion protection efficiency (% P.E.) was determined from the measured i_{corr} (corrosion current densities with blank mild steel electrode (i^0_{corr}) without coatings and corrosion current densities with a mild steel electrode coated with polymer coated (i^c_{corr}) values by using the following relationship;

$$P.E.(\%) = \frac{i^0_{corr} - i^c_{corr}}{i^0_{corr}} \times 100 \tag{3}$$

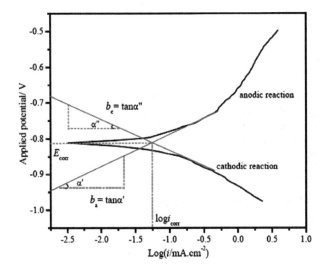

Figure 1. Schematic polarization curve showing Tafel extrapolation.

2.2. Weight loss method

The weight loss methods have also been performed for corrosion study. Polymer coated mild steel specimens of dimension 4 x 3.5 cm^2 have been tested for same span of time by immersing the samples in 1.0 M HCl or aqueous 3.5 % NaCl solution for 60 days. The uncoated and polymers coated mild steel specimens were weighed in an electronic balance with an accuracy of 0.1 mg. before immersion in saline medium. After the 60 days of immersion the mild steel specimens were withdrawn from the tested solution, washed thoroughly with distilled water followed by acetone and dried with air, then weighed again. The performance of the coating was examined visually and through calculation of the weight loss. Weight loss (W.L) is expressed as the loss in the weight per unit area or per unit area per unit time (g cm^{-2} h^{-1}) as follows:

$$W.L = \frac{w_0 - w_1}{a.t} \qquad (4)$$

where, w_0 = initial weight of the sample before immersion (g); w_1 = weight of the sample after immersion (mg); a= surface area (cm^2) of specimen; t = end time (h) of each experiment. If we introduce the density of metal; d (g/cm^3), the loss in the thickness of metal per unit time can be calculated. Corrosion rate (C.R) in mm/year can also be calculated by weight loss method as follows:

$$C.R(mm / year) = \frac{(w_0 - w_1) \times 87.6}{a.t.d} \qquad (5)$$

2.3. Surface study

Surface studies comprise the analysis of surface of metals before and after corrosion in order to estimate the rate as well as mechanism of corrosion. Techniques like scanning electron microscopy and electron probe micro analysis are used to study the structure, chemical composition of corrosion product formed onto metal surface. Other techniques like atomic force microscopy and ellipsometry are used to study the surface of metals with and without corrosion.

3. Synthesis of the SiO$_2$ nanoparticles

The syntheses of mono disperse uniform- sized SiO$_2$ nanoparticles were carried out by using ammonia as catalyst and ethanol as solvent. Hydrolysis method of tetra-ethylorthosilicates (TEOS) was used for synthesizing SiO$_2$ nanoparticles. Aqueous ammonia (0.1M) was added to a solution containing ethanol (1.0 M) and 20 ml of deionized water which was stirred for 1 h then 0.05M TEOS was added and again stirred for 1 h at room temperature. Appearance of white turbid suspension indicating the formation of silicon dioxide, this suspension was retrieved by centrifugation and further calcination at 823 K for 6 hours.

4. Preparation of PANI/SiO$_2$ composites:

4.1. Chemical oxidative polymerization

PANI/SiO$_2$ composites were prepared by *in situ* chemical oxidative polymerization of aniline using APS as an oxidant. Weight ratio of aniline and SiO$_2$ was taken as 1:1 for preparation of PANI/SiO$_2$ nanocomposites. Aniline was adsorbed on SiO$_2$ particles and 0.2 M phosphoric acid/0.2 M perfluro octanoic acid (PFOA) was added in this solution. Polymerization was initiated by drop wise addition of ammonium persulphate solution (APS) (0.1 M, (NH)$_4$S$_2$O$_8$ in distilled water). The polymerization was carried out at a temperature of 0-3 C for a period of 4-6 h. The synthesized polymer composite was isolated from reaction mixture by filtration and washed with distilled water to remove oxidant and oligomers and followed by drying in the vacuum oven at 60°C.

4.2. Electro-chemical polymerization

The electrochemical polymerization of 0.1 M aniline and aniline in the presence of SiO$_2$ in 0.2 M H$_3$PO$_4$ /0.2 M PFOA was carried out between −0.20 to 1.5 V on platinum electrode vs. Ag/AgCl reference electrode. The polymer film growth was studied by sweeping the potential between −0.20 to 1.5V on Pt electrode at a scan rate of 20 mV/s. Prior to polymerization, the solution was deoxygenated by passing argon gas through the reaction solution for 30 min. Peak potential values of the corresponding polymer and PANI-SiO$_2$ composites were recorded in 0.2 M H$_3$PO$_4$ /0.2 M PFOA medium.

5. Preparation of PANI/SiO$_2$ composites coated mild steel

Mild steel coated with PANI/SiO$_2$ nanocomposites electrode of dimension 1 cm x 1 cm were employed to carry out the corrosion studies. Surface treatments were applied on the samples including the cleaning of the electrode was carried out by 1/0, 2/0, 3/0 and 4/0 grade emery papers. The electrodes were then thoroughly cleaned with acetone and trichloroethylene to remove any impurities on the surface. The powder polymer was mixed with epoxy formulation in various proportions ranging from 1.0 % to 6.0 wt. %. The polymer-epoxy powder coating was applied to a thickness of 45 ± 3 μm using an electrostatic spray gun. After obtaining uniform coverage of the powder the powder-coated panels were placed in air drying oven for curing at 140°C for 20 min. The adhesion of the coating was tested by tape test as per ASTM D3359-02 and found to pass the test.

6. Characterization of PANI-SiO$_2$ nanocomposites

6.1. Electrochemical behaviour

Figure 2 shows the electrochemical growth behaviour of aniline and aniline-SiO$_2$ in 0.2 M H$_3$PO$_4$. The polymer film growth was studied by sweeping the potential between -0.20 and 1.5 V on Pt electrode at a scan rate of 20 mV/s. Peak potential values of the corresponding PANI and PSC were recorded in H$_3$PO$_4$ medium. First anodic peak (oxidation peak) corresponds to the oxidation of monomer. During the first reverse sweep, a reduction peak appears which shows that the formation of oligomers and polymer on electrode surface as shown in Figure 2.

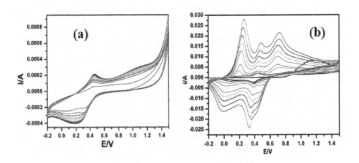

Figure 2. Electrochemical growth behaviour of (a) 0.1M aniline in 0.2M H$_3$PO$_4$ medium and (b) aniline- SiO$_2$ in 0.2M H$_3$PO$_4$ medium in potential range between -0.2 V to 1.5 V vs. Ag/AgCl at scan rate of 20 mV/s.

After the first scan, well defined oxidation and reduction peaks of polymers between 0.2 and 0.6 V vs. Ag/AgCl appeared. The current values of each oxidation and reduction peaks are greater than that of a previous cycle which indicate the built up of an electro active polymeric material on the electrode surface.

Figure 3 shows the cyclic voltammogram of PANI-SiO$_2$ composite and the inset figure shows the cyclic voltammogram of PANI in H$_3$PO$_4$ medium. We have observed quite interesting observation when we recorded the cyclic voltammogram of aniline in phosphoric acid medium and when SiO$_2$ nanoparticles were incorporated in the monomer matrix. On recording the cyclic voltammogram of aniline in H$_3$PO$_4$ medium, it was observed that anodic peak potential is observed at 0.456 V.

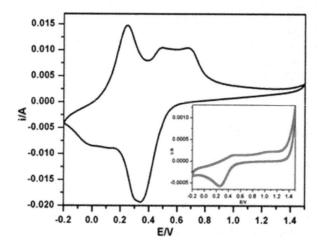

Figure 3. Cyclic voltammogram of PSC and inset shows the cyclic voltammogram of PANI in H$_3$PO$_4$ medium at a scan rate of 20 mV/s.

However, in the presence of SiO$_2$ matrix, these peaks appeared at 0.247 V. The reason for getting these deviations in H$_3$PO$_4$ medium is that phosphoric acid is a weak protonic acid whose pKa1 value is 2.21 which results in shifting of peak potential value to 0.456 V. Conventionally, in strong acidic medium like 1.0 M HCl medium, these values of peak potential for PANI are observed at 0.1 V. However, when the CV was recorded for aniline in the presence of SiO$_2$ and H$_3$PO$_4$ medium, there is a possibility that protons from phosphoric acid medium might have protonated SiO$_2$ resulting in generation of well-defined peaks as observed in the growth behavior of CV resulting in observing peak potential values at 0.247 V (Fal). These experiments were repeated by us number of times and each time this type of cyclic voltammogram was observed which has led us to draw the above conclusion. Protons from phosphoric acid might have led to the formation of protonated silica thereby shifting of peak potential values which might have enhanced the electropolymerization of aniline.

6.2. FTIR spectra

Figure 4 shows the FTIR spectra of SiO_2, PANI and PANI-SiO_2. PANI showed the main characteristics bands at 1565 and 1475 cm^{-1} attributed to the stretching mode of C=N and C=C, the bands at 1292 and 1245 cm^{-1} indicating the C–N stretching mode of benzenoid ring and the band at 1117 - 1109 cm^{-1} is assigned to a plane bending vibration of C–H mode which is found during protonation [60]. The FTIR spectra of SiO_2 indicated that the characteristic peak at 1081 cm^{-1} and 807 cm^{-1} are assigned to the stretching and bending vibration of Si–O–Si respectively. By comparing the peaks of PANI and PSC, it was observed that some peaks of PSC were shifted due to the presence of SiO_2 particles in polymer matrix. For example, the peaks at 1565 cm^{-1}, 1475 cm^{-1}, 1292 cm^{-1}, and 1245 cm^{-1} shifted to higher wavenumbers, and the bending vibration of Si–O–Si peak at 1056 cm^{-1} shifted to the lower wavenumbers. These changes also indicate that an interaction exists between PANI molecule and SiO_2 particles. These peaks were also observed in PSC indicating the interaction of SiO_2 particles in polyaniline chain.

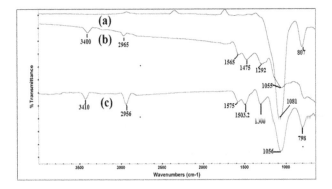

Figure 4. FTIR spectra of (a) SiO_2 (b) PANI and PSC.

6.3. Thermogravimetric analysis

Figure 5 shows the thermo-gravimetric curves (TG) of pure SiO_2, PANI and their composites. The materials were heated from 25 to 800°C under a constant heating rate of 10°C/min and in the inert atmosphere of nitrogen gas (60 ml/min). The SiO_2 particle has excellent thermal stability up to 800°C and weight loss was only 0.15 %. The TGA curve of PSC indicated, first weight loss at 110°C may be attributed to the loss of water and other volatiles species. The weight loss in the second step at about 280°C involves the loss of phosphate ions as well as onset of degradation of polyaniline backbone. The increasing SiO_2 content slightly affects the decomposition temperature (DT) which increases from 280°C (PANI) to 295°C (PSC). The third weight loss step between 300 to 550°C can be ascribed to the complete degradation of dopant as well as polymeric backbone. The composites show little weight loss between the 500-800°C and the residue remaining in this region gives an approximate estimate of filler content. Therefore, the final weight of SiO_2 incorporated in polymer was found to 21 %. The results indicate that actually incorporated SiO_2 fraction is less than the ratio of aniline: SiO_2 taken in the initial reaction mass.

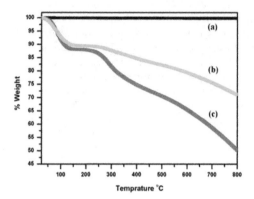

Figure 5. Thermal gravimetric analysis of SiO_2, PANI and (c) PSC doped with H_3PO_4.

The TGA data clarify that these composites are thermally stability up to 295°C, which envisages them as a good candidate for melt blending with conventional thermoplastics like polyethylene, polypropylene, polystyrene etc.

6.4. UV-Visible spectra

Figure 6 shows the UV absorption spectra of polyaniline and its composite with SiO_2. We have measured the UV absorption spectra of polymer using dimethyl sulfoxide (DMSO) as a solvent from 250 to1100 nm.

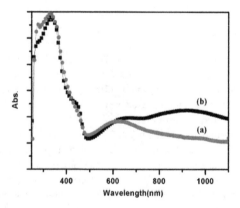

Figure 6. UV–Visible spectra of (a) PANI & (b) PSC in DMSO.

The UV-visible absorption data indicates that λ_{max} values in case of polyaniline doped with o-phosphoric acid medium in DMSO solvent lies at 326nm, 431nm and 619 nm whereas in PSC composite these values lies at 336 nm, 447 nm and 654 nm. In case of polyaniline, π-π^* transition [61, 62] occurs at 326 nm whereas in case of PSC composite, this transition value lies at 336 nm.This indicates that the addition of SiO_2 particles absorbed in aniline matrix and on polymerization in o-phosphoric acid medium may have caused some interactions with polymer matrix resulting in shifting of bands from 326 nm to 336 nm. This is the reason of shifting of polaronic bands which also shows a shift from 431 nm to 447 nm and 619 nm to 654 nm.

7. Anticorrosive properties of PANI and PSC coated mild steel in 1.0 M HCl medium.

7.1. Chronoamperometry method

Figure 7 shows the chronoamperometric response of uncoated, epoxy coated, PANI and PSC coated mild steel sample in 1.0 M HCl. After the samples reached a stable OCP (open circuit potential), a potential in the range of 1.2 V vs SCE was applied and current was recorded as a function of time. It was observed that the current of PANI and PSC coated mild steel sample remained at a very small value as compared with the uncoated mild steel electrode indicating the good protective properties by these polymers coating. Moreover, it has been observed that the current density value of PSC coated mild steel was lower than that of PANI-coated mild steel sample. The decrease in current density with increasing amount of PSC material in epoxy resin. Hence, chronoamperometric test results showed that mild steel coated with PANI/SiO_2 composites shows the higher corrosion protection performance as compared to PANI coated mild steel samples. This statement was further confirmed by other corrosion test methods like Tafel extrapolation and weight loss methods.

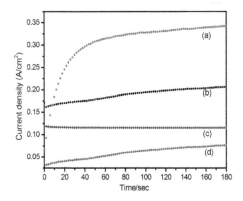

Figure 7. Chronoamperometric response of (a) uncoated (b) epoxy coated (c) PANI and (d) PSC coated mild steel sample in 1.0 M HCl.

7.2. Tafel extrapolation method

Tafel polarization behaviour of mild steel in 1.0 M HCl with uncoated, epoxy coated, PANI and PSC coated mild steel are shown in the Figure 8& Figure 9.

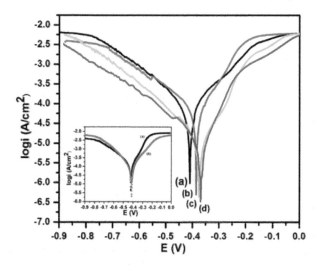

Figure 8. Tafel curves of PANI coated mild steel electrode with different loading level of PANI in epoxy resin (a) 1.5% (b) 3.0% (c) 4.5% (d) 6.0% whereas the inset shows (a) blank mild steel electrode and (b) epoxy coated mild steel electrode in 1.0 M HCl.

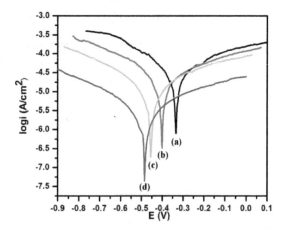

Figure 9. Tafel curves of PSC coated mild steel electrode with different loading level of PSC in epoxy resin (a) 1.5% (b) 3% (c) 4.5% (d) 6.0%.

The corrosion kinetic parameters derived from these curves are given in the Table 1. As shown in Table 1, PSC coated mild steel sample showed a remarkable current shift from 132 μA to 0.09 μA versus Ag/AgCl in the corrosion current (i_{corr}), relative to the value of the un-coated mild steel.

The significant reduction in the corrosion current density (i_{corr}) in polymer coated mild steel indicated the effective corrosion protection performance of these polymers. The corrosion current values (i_{corr}) were found to be decreased from 132 μA/cm² for uncoated mild steel sample to 107.6 μA/cm² for epoxy coated mild steel sample to 0.09 μA /cm² for PSC coated mild steel samples.

The corrosion current values (i_{corr}) decreased with increasing the concentration of PSC in ep-oxy resin. i_{corr} value decreased from 15.4 μA/cm² at 1.5 wt. % to 0.09 μA/cm² at 6.0 wt. % loading of conducting material in epoxy resin as shown in Figure 9. While PANI coated mild steel showed the i_{corr} in the range of 10.9 μA/cm² at 6.0 % loading as shown in Figure 8.

Sample name	Loading level of polymer (%)	I_{corr} (μA/cm²)	Corrosion rate (mm/year)	Protection efficiency (%)
Blank mild steel	-	132.0	1.54	--
Epoxy coated mild steel	--	107.6	1.26	18.48
PANI	1.5	98.8	1.15	25.15
	3.0	75.4	0.88	42.88
	4.5	20.5	0.23	84.47
	6	10.9	0.13	91.74
PANI-SiO₂ Composite (PSC)	1.5	15.4	0.18	88.33
	3.0	9.09	0.11	93.11
	4.5	5.12	0.05	96.12
	6	0.09	0.0011	99.93

Table 1. Tafel parameters for corrosion of mild steel in 1.0 M HCl with different loading level of PANI & PSC in epoxy resin.

The corrosion protection efficiency calculated from Tafel parameter revealed that PANI coated mild steel showed 25% protection efficiency at 1.5 wt.% loading while PSC coated mild steel showed 88 % P.E at the same loading level. Up to 99.93 % protection efficiency have been achieved by using 6.0 wt.% loading of PSC in epoxy resin.

7.3. SEM studies of uncoated and coated mild steel observed by weight loss method

The scanning electron micrographs of SiO₂ particles, PANI and PSC are shown in Figure 10. SiO₂ particles showed spherical shaped morphology and PANI showed globular morpholo-gy. Figure 10 b shows the TEM image of of SiO₂ particles, which indicates the dimension of

SiO_2 particles, was found to be 90-100 nm. Morphology of PSC indicates the incorporation of SiO_2 particles in PANI matrix. SEM image of PSC revealed that the entrapment of SiO_2 particles in the globular space of PANI matrix during in situ polymerisation of polyaniline.

Figure 10. SEM micrographs powder sample of (a) SiO_2 (c) PANI (d) PSC and (b) TEM image of SiO_2 particles.

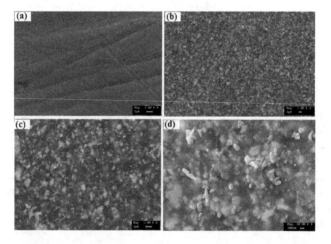

Figure 11. SEM micrographs of (a) blank mild steel electrode (b) blank epoxy resin coated electrode (c) PANI coated (d) PSC coated electrode before immersion in 1.0M HCl

SEM images of uncoated, epoxy coated and polymer coated samples before and after the immersion test of 60 days have been shown in Figure 11 & Figure 12. These images clearly show the formation of large pits on the surface of mild steel after immersion. These pits and cracks were developed during the corrosion of mild steel in acidic medium. In the case PANI coated sample, few pits still appeared on mild steel surface. While, PSC coated mild steel samples did not show any cracks and pits on the metal surface. No detachment of coating from mild steel substrate was also observed after the immersion of these samples in 1.0 M HCl medium for 60 days of immersion indicating strong adherence of PSC composite to the mild steel substrate and it is resistant to corrosion in aqueous 1.0 M HCl solution as shown in Figure 13.

When epoxy coated mild steel sample was immersed in the acidic medium for 60 days, detachment of coating from mild steel substrate have been observed. The pits were also appeared on the metal surface as shown in Figure 12 b.

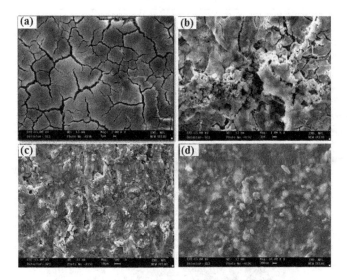

Figure 12. SEM micrographs of (a) blank mild steel electrode (b) blank epoxy resin coated electrode (c) PANI coated (d) PSC coated electrode after immersion in 1.0 M HCl for 60 days.

It was found that the PSC content has a great influence on the anticorrosive performance of the coating. The corrosion protection effect of PSC coated mild steel sample improved slowly when PSC content in epoxy formulation increases from 1.5 to 3.0 wt. % Afterward, an excellent corrosion protection effect appears at 6.0 wt. % loading of PSC content in epoxy resin.

Corrosion rates (C.R) in mm/year have also been calculated by weight loss method and the values have been given in Table 2. It was observed that the corrosion rate was highest for uncoated mild steel in HCl medium. After 60 days of immersion, the C.R value of uncoated mild steel was found to 7.25 mm/year. Epoxy and PANI coated samples showed C.R. value of 6.37 mm/year and 1.90 mm/year respectively. While in the case of PSC coated sample in 6.0 % loading, C.R. value reduced to 0.73 mm/year.

Sample name	Loading level of polymer (%)	Initial Weight (mg)	Weight After immersion in HCl for 60 days (mg)	Weight loss (mg)	Weight loss (%)	C.R (mm/year)	Protection Efficiency (%)
Blank mild steel	-	30354.6	17256.30	13098.30	43.15	7.25	-
Blank epoxy	0	31113.4	19598.61	11514.79	37.00	6.37	12.10
PANI	1.5	31717.2	21518.09	10199.11	32.15	5.64	22.13
	3.0	31601.2	23698.65	7902.55	25.00	4.37	39.67
	4.5	31289.1	27021.37	4267.73	13.63	2.36	67.42
	6	30454.0	27018.20	3435.80	11.28	1.90	73.77
PANI-SiO2 Composite	1.5	32487.2	29606.00	881.20	8.86	1.59	78.00
	3.0	32891.5	30397.90	2493.60	7.58	1.38	80.96
	4.5	31856.7	30031.30	1825.40	5.73	1.01	86.06
	6	31773.7	30454.20	1319.50	4.15	0.73	89.93

Table 2. Weight loss parameter of uncoated and coated mild steel samples after immersion test in 1.0 M HCl for 60 days.

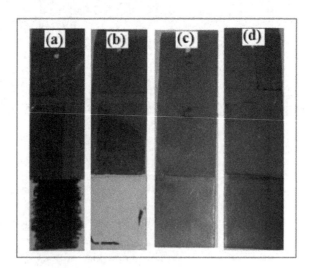

Figure 13. Photographs of (a) blank mild steel electrode (b) blank epoxy resin coated electrode (c) PANI coated (d) PSC coated electrode after immersion in 1.0 M HCl for 60 days.

8. Characterization of hydrophobic PANI - SiO₂ Nanocomposites (HPSC)

8.1. Electrochemical behaviour

Figure 14 shows the electrochemical growth behavior of aniline and aniline-SiO$_2$ in 0.2 M PFOA solution. Electrochemical polymeriztion was carried out at 0.9 V on platinum electrode vs Ag/AgCl reference electrode. The polymer film growth was studied by sweeping the potential between -0.20 and 0.9 V on Pt electrode at a scan rate of 20 mV/s.

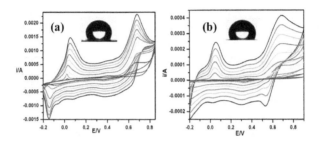

Figure 14. Electrochemical growth behaviour of (a) 0.1M aniline in 0.2M PFOA medium and (b) aniline- SiO$_2$ in 0.2M PFOA medium in potential range between -0.2 V to 0.9 V vs. Ag/AgCl at scan rate of 20 mV/s.

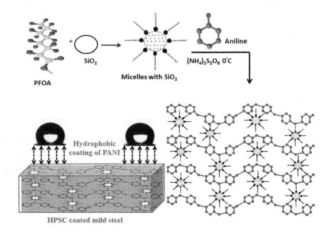

Figure 15. Schematic representation of formation of HPSC coating onto mild steel surface.

Peak potential values of the corresponding PANI and HPSC were recorded in PFOA medium. First anodic peak (oxidation peak) corresponds to the oxidation of monomer. The intensity of this peak gradually decreases with subsequent scans. During the first reverse sweep, a reduction peak appears which shows that the formation of oligomers and polymer on electrode surface as shown in Figure 14. Figure 15 shows the Schematic representation of formation of HPSC coating onto mild steel surface. After the first scan, well defined oxidation and reduction peaks of polymers between 0.2 and 0.6 V vs. Ag/AgCl appeared. The current values of each oxidation and reduction peaks are greater than that of a previous cycle which indicate the built up of an electroactive polymeric material on the electrode surface. Moreover, it was observed that current value of PANI film was found to be higher than that of HPSC film which revealed higher conductivity of PANI as compare to HPSC coating on electrode surface.

Cyclic voltammogram of HPSC and PANI in PFOA medium indicates that the first peak potential value of PANI in PFOA medium lies at 0.15 V. Incorporation of SiO_2 particle in PANI, the first peak potential value shifted from 0.15 V to 0.041 V vs Ag/AgCl as shown in Figure 16.

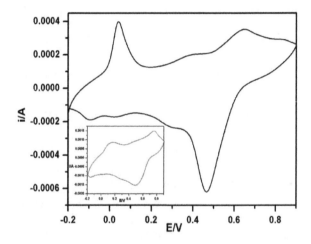

Figure 16. Cyclic voltammogram of HPSC and inset figure shows the cyclic voltammogram of PANI in PFOA medium at a scan rate of 20 mV/s.

This implies that the polymerization of aniline leads to larger peak potential shift as compared to the aniline in presence of SiO_2 which indicates the presence of SiO_2 particles in the polymer chain induce some change in configurations along the polymer backbone which is responsible for the negative shift in the oxidation potential.

8.2. FTIR spectra

The Figure 17 shows the FTIR spectra of SiO_2, PANI and HPSC. PANI showed the main characteristics bands at 1554 and 1438 -1440 cm^{-1} attributed to the stretching mode of C=N and C=C, the bands at 1250 cm^{-1} indicating the C–N stretching mode of benzenoid ring.

The FTIR spectra of SiO_2 indicated that the characteristic peak at 1081 cm^{-1} and 807 cm^{-1} are assigned to the stretching and bending vibration of Si–O–Si respectively. These peaks were also observed in HPSC indicating the interaction of SiO$_2$ particles in polyaniline chain. HPSC and PANI showed a characteristic strong peak at 1738 cm^{-1} due to C=O stretching mode and peak at 1365 cm^{-1} due to C-F stretching mode [61], which indicates the interaction of PFOA dopant in the polymer chain.

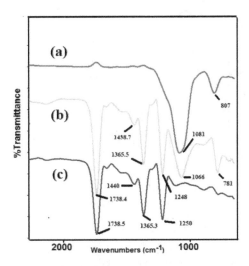

Figure 17. FTIR spectra of (a) SiO$_2$ (b) PANI and (c) HPSC.

8.3. Wettability of HPSC coating

The surface wettability was measured by static contact angle measurements with water (γ=72.8 mN/m) to determine surface hydrophobicity. The drop volume used for the measurements was 2.0 µL. The PANI-SiO$_2$ nanocomposite (HPSC) coated electrodes exhibited hydrophobic properties with static water contact angle of about 115° as shown in Figure 14.

9. Anticorrosive properties of HPSC coated mild steel in 3.5 % NaCl solution

9.1. Tafel Extrapoaltion method

Tafel polarization behaviour of mild steel in 3.5 % NaCl solution with uncoated, epoxy coated, PANI and HPSC coated mild steel are shown in the Figure 18& Figure 19. The corrosion kinetic parameters derived from these curves are given in the Table 3. As shown in Table 3. HPSC coated mild steel sample showed a remarkable current density shift from 106.5

$\mu A/cm^2$ to 4.36 $\mu A /cm^2$ versus Ag/AgCl in the corrosion current (I_{corr}), relative to the value of the uncoated mild steel.

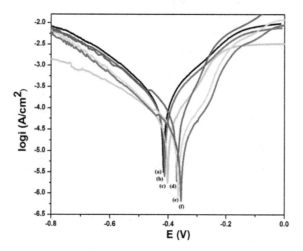

Figure 18. Tafel curves of uncoated and polymer coated mild steel electrode in 3.5 % NaCl solution (a) blank electrode (b) epoxy coated mild steel (c) PANI coated mild steel at 1.5 wt. % loading (d) 3.0 wt. % loading (e) 4.5 wt.% loading and (f) 6.0 wt. % loading of PANI in epoxy resin.

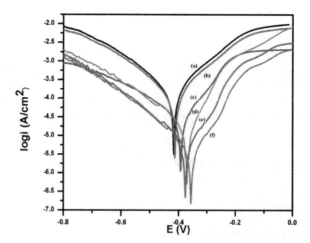

Figure 19. Tafel curves of (a) blank mild steel (b) epoxy coated mild steel (c) HPSC coated mild steel at 1.5 wt. % loading (d) 3.0 wt. % loading (e) 4.5 wt.% loading (f) 6.0 wt. % in 3.5 % NaCl solution.

The significant reduction in the corrosion current density (i_{corr}) in polymer coated mild steel indicated the effective corrosion protection performance of these polymers. The corrosion current values (i_{corr}) found to be decreased from 106.5 µA/cm² for uncoated mild steel sample to 98 µA/cm² for epoxy coated mild steel sample to 4.36 µA /cm² for HPSC coated mild steel samples. The corrosion current values (i_{corr}) decreased with increasing the concentration of HPSC in epoxy resin. i_{corr} value decreased from 32.6 µA/cm² at 1.5 wt. % to 4.36 µA/cm² at 6.0 wt. % loading of conducting material in epoxy resin.

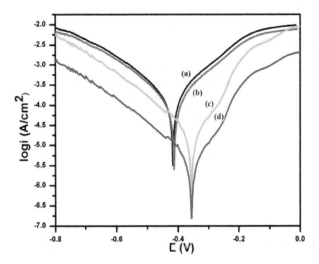

Figure 20. Tafel curves of (a) blank electrode (b) epoxy coated (c) PANI coated mild steel at 6.0 wt. % loading and (d) HPSC coated mild steel at 6.0 wt. % loading in 3.5 % NaCl solution.

Sample name	Loading level of polymer (%)	I_{corr} (µA/cm²)	Corrosion rate (mm/year)	Protection efficiency (%)
Blank mild steel	-	106.5	1.51	0
Epoxy coated mild steel	0	98.0	1.39	7.9
PANI	1.5	64.5	0.91	39.43
	3.0	61.6	0.87	42.16
	4.5	30.2	0.42	71.6
	6	20.4	0.28	80.84
HPSC	1.5	32.6	0.45	69.39
	3.0	12.8	0.17	87.98
	4.5	6.16	0.08	94.21
	6	4.36	0.06	96.0

Table 3. Tafel parameters for corrosion of mild steel in 3.5% NaCl with different loading level of PANI & HPSC in epoxy resin.

While PANI coated mild steel showed the i_{corr} in the range of 20.4 µA/cm^2 at 6.0 % loading. The corrosion protection efficiency calculated from Tafel parameter revealed that PANI coated mild steel showed the protection efficiency 39.4 % at 1.5 wt.% loading while HPSC coated mild steel showed 69.4 % P.E at the same loading level.

Up to 96 % protection efficiency have been achieved by using 6.0 wt. % HPSC in epoxy resin. While in case of PANI, only 80.84% protection has been achieved at 6.0 wt. % loading, as shown in Figure 20.

9.2. Weight loss method

Table 4 shows the values of the weight loss from mild steel samples during the immersion test. The results revealed that HPSC coated samples were more protectable to mild steel than that of only PANI coated samples in same immersion time. After the immersion of coated and uncoated samples in 3.5% NaCl solution for 60 days, it was observed that uncoated and epoxy coated samples showed the maximum weight loss of 34.18 % and 29.11 % respectively

Sample name	Loading level of polymer (%)	Initial Weight (before immersion) (mg)	Final Weight After immersion (mg)	Weight loss (mg)	Weight loss (%)	C.R (mm/ year)	P.E (%)
Blank mild steel	-	32632.1	21478.45	11153.7	34.18	4.18	0
Blank epoxy	0	30918.7	21918.27	9000.4	29.11	3.56	14.83
PANI	1.5	32798.2	24913.5	7884.7	24.04	2.94	29.67
	3.0	30416.2	23502.6	6913.6	22.73	2.78	33.49
	4.5	32678.1	26642.5	6035.6	18.47	2.26	45.93
	6	31567.2	28369.4	3197.8	10.13	1.24	70.33
HPSC	1.5	31494.2	25541.8	5952.4	18.9	2.32	44.50
	3.0	30566.5	25935.7	4630.8	15.15	1.86	55.50
	4.5	32929.7	29926.5	3003.2	9.12	1.12	73.20
	6	30804.7	30102.4	702.3	2.28	0.28	93.30

Table 4. Weight loss parameter of uncoated and coated mild steel samples after immersion test in 3.5 % NaCl for 60 days.

PANI coated mild steel showed the weight loss up to 10.13 % at 6.0 wt. % loading while HPSC coated samples at the same loading level showed negligible weight loss (i.e < 3 %) after 60 days of immersion in 3.5 % NaCl medium. Corrosion rate (C.R) in mm/year have also been calculated by weight loss method. It was observed that the corrosion rate was highest for uncoated mild steel in 3.5 % NaCl medium. After 60 days of immersion, the C.R value of uncoated mild steel was found to 4.18 mm/year. Epoxy and PANI coated samples showed C.R. value of 3.56 mm/year and 1.24 mm/year respectively. While in the case of HPSC coated sample in 6 % loading, C.R. value reduced to 0.28 mm/year.

The appearance of the uncoated and HPSC coated mild steel samples after exposure to salt spray fog for 35 days is shown in Figure 21.

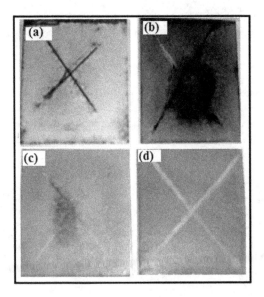

Figure 21. Photograph of (a) epoxy coated (b) PANI (at 6 wt.% loading) coated (c) HPSC (at 1.5 wt. % loading) and (d) HPSC (6.0 wt. % loading mild steel after 35 days of exposure to salt spray test.

It was observed that epoxy coated and PANI coated (at 6.0 wt. % loading) mild steel have more corrosion extended area from the scribes as shown in Figure 21 (a) and 21(b) while HPSC coated mid steel (at 1.0 wt. % loading) showed less corrosion extended area as compared to epoxy and PANI coated mid steel as shown in Figure 21(c). However, HPSC containing coating sample (at 6.0 wt.% loading) were found to be free from rust and blister as shown in Figure 21 (d). Moreover, there was no spreading of rust along the scribed areas.

9.3. SEM studies of uncoated and coated mild steel before and after immersion test

It was observed that SiO_2 particles showed spherical shaped morphology as shown in Figure 10 a. The scanning electron micrographs of PANI and HPSC are shown in Figure 22. PANI doped with PFOA showed uniform net like morphology as shown in Figure 22(a). Morphology of HPSC was entirely different with incorporation of SiO_2 particles in PANI matrix during polymerisation. SEM image of HPSC revealed that the entrapment of SiO_2 particles in the globular space of PANI matrix during in-situ polymerisation of polyaniline as shown in Figure 22 (b).

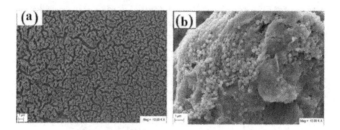

Figure 22. SEM micrographs of powder sample of (a) PFOA doped PANI, and (b) HPSC.

SEM images of uncoated, epoxy coated and polymer coated samples before and after the immersion test of 60 days have been shown in Figure 23 and 24 respectively. These images clearly show the formation of large pits on the surface of mild steel after immersion. These pits and cracks were developed during the corrosion of mild steel in NaCl medium.

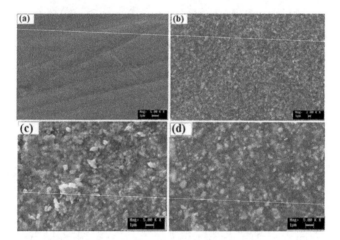

Figure 23. SEM micrographs of mid steel electrode (a) blank (b) epoxy coated (c) PANI coated (d) HPSC coated before immersion.

When epoxy coated mild steel sample was immersed in the acidic medium for 60 days, detachment of coating from mild steel substrate have been observed. The pits were also appeared on the metal surface as shown in Figure 24. In the case PANI coated sample, few pits still appeared on mild steel surface. While, HPSC coated mild steel samples did not show any cracks and pits on the metal surface. No detachment of coating from mild steel substrate was also observed after the immersion of these samples in 3.5 % NaCl medium for 60 days of immersion indicating strong adherence of HPSC composite to the mild steel substrate and it is resistant to corrosion in aqueous 3.5 % NaCl medium.

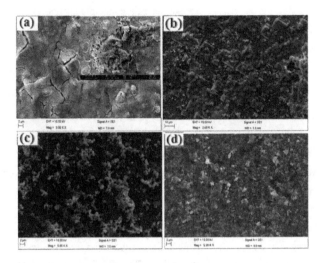

Figure 24. SEM micrographs of mid steel electrode (a) blank (b) epoxy coated (c) PANI coated (d) HPSC coated after immersion.

It was found that the HPSC content has a great influence on the anticorrosive performance of the coating. The corrosion protection effect of HPSC coated mild steel sample improved slowly when HPSC content in epoxy formulation increases from 1.5 to 3.0 wt. % Afterward, an excellent corrosion protection effect appears at 6.0 wt. % loading of HPSC content in epoxy resin.

9.4. Mechanism of corrosion protection of PANI-SiO$_2$ nanocomposites

The corrosion studies show that the PANI-SiO$_2$ nanocomposites containing coating showed better corrosion protection as compared to PANI coating which may be due to the redox property and uniform distribution of PANI in the coating containing PANI-SiO$_2$ nano-composites. Earlier studies [63-64] have shown that the redox property of PANI coating on metal surface plays an important role to protect the metal by passivating the pin holes. Corrosion protection of metals occurs via reduction of PANI–Emeraldine salt (PANI–ES) to PANI–Leucosalt (PANI–LS) with the concomitant release of phosphate dopant [19]. Phosphate ions help to form passive film on mild steel at the defect. PANI–LS is assumed to undergo a subsequent re-oxidation by dissolved oxygen to PANI–ES. Due to this cyclic reaction, the coating containing PANI is able to offer higher corrosion protection. However, in case of PANI-SiO$_2$ nano composites containing coating, these composites have a dual protection mechanism; forming a passive layer and simultaneously acting as a physical barrier to avoid chloride ion penetration. Moreover, it acts as a barrier between metal surface and corrosive environment. Entrance of water and corrosive ions on the metal surface causes the defects in the paint coating and therefore the protective property of the coating is decreased. Due to uniform distribution of PANI, the possibility of forming uniform passive layer on the mild steel surface is more since PANI has been shown to protect the mild steel surface by passive film formation. Furthermore, powder coating technique also plays an important role for achieving high quality, durable and good anticorrosive coatings.

Corrosion protection property of these coating may also be attributed to the PSC/HPSC content in epoxy resin which can react with epoxy to form highly adherent, dense and non porous polymer film on the mild steel surface. On the other hand, presence of SiO_2 nanoparticles entrapped in PANI chain provide the reinforcement to PANI chain which reduce the degradation of polymer chain in corrosive condition.

10. Antistatic performance of the conducting polymers nanocomposites based on nanotubes of poly (aniline-co-1-amino-2-naphthol-4-sulphonic acid)/LDPE composites

10.1. Introduction

Electrostatic charge dissipation has become an important issue within the electronic components such as data storage devices, chips carriers and computer internals. Antistatic protection is also required for parts where relative motion between dissimilar materials occurs [65] like weaving machine arms, airplane tyres etc. Conventional polymers commonly being used for packaging of various electronic equipments but due to their inherent electrical insulating nature, these polymers failed to dissipate the static or electrostatic charge. The generation of static electricity on the materials leads to a variety of problems in manufacturing and consumer use. Moreover, electronic components are susceptible to damage from electrostatic discharge. Thus the challenge is to convert inherently insulating thermoplastic to a product that would provide an effective antistatic material. Various attempts have been made to achieve the antistatic polymers with retained mechanical properties such as addition of antistatic agents [66], conducting additives [67] and fillers like carbon powder [68] and carbon nanotubes [69] etc. The electrical conductivity of the polymeric material depends on the amount, type and shape of the conducting filler [70]. According to electronic industries association (EIA) standards, in ESD protected environments, the optimal surface conductivity should be in the range of 10^{-6} to 10^{-10} S/cm. However the functioning of antistatic agents is critically dependent on the relative humidity [71] whereas the metal and carbon filled materials suffer from the problems like bleeding and poor dispersion [72]. Moreover, it has been observed that carbon black loaded static controlling materials usually contain 15-20 % carbon black. The addition of carbon black at higher concentrations showed a negative effect on the proccessability of the compound and mechanical properties such as increase in melt viscosity and decrease in impact resistance. Use of conducting blends and composites with conventional polymers as an electrostatic charge dissipative material is one of the promising application of conducting polymers which combines the mechanical properties of conventional polymers and electrical properties of conducting polymers. Polyaniline is one of the most promising intrinsically conducting polymer (ICPs) because of its good environmental stability and high electrical conductivity, which can be reversibly controlled by a change in the oxidation state and protonation of the imine nitrogen groups. Blending of polyaniline with conventional polymers like polypropylene, ABS, LDPE etc. can also be used to improve the proccessability of polyaniline creating new materials with specific properties for the desired application at low cost that can also be used for different applications like electromagnetic shielding and corrosion prevention where conductivity, proccessability and mechanical properties of the materials are of the primary importance. Hence desired properties of conducting polymers can be enhanced by mixing it with a polymer that has good mechanical properties and the unique combination of electrical and mechanical properties of conducting copolymers blends with insulating polymers seems to have great potential for their use in many applications [73-75].

11. Antistatic measurements

Antistatic or electrostatic charge dissipative performances of blends of conducting polymers were measured by Static decay meter, John Chubb Instrument and Static Charge Meter. The detailed method is given below.

11.1. Static decay meter

Static decay meter is very useful device for measurement of static decay time of the conducting polymer blends in the form of injection moulded sheets and blown film. The samples of conducting copolymer blends (LDPE/conducting copolymer) were cut in to the 15 x 15 cm^2 blown film and was used for measurement of static decay time on Static Decay Meter by measuring the time on applying a positive voltage of 5000 V and recording the decay time on going down to 500 V. Similarly the static decay time was measured by applying a negative voltage of 5000 V. Here, these measurements were carried out on Static Decay Meter; model 406D, Electro-tech System, Inc., USA. The model 406D Static Decay Meter is designed to test the static dissipative characteristics of material by measuring the time required for charge test sample to discharge to a known, predetermined cut-off level. Three manually selected cut-off threshold at 50%, 10% and 1% of full charge are provided and samples are charged by an adjustable 0 to ±5kV high voltage power supply.

11.2. John Chubb Instrument

John Chubb Instrument (JCI 155 v5) charge decay test unit is a compact instrument for easy and direct measurement of the ability of materials to dissipate static electricity and to assess whether significant voltage will arise from practical amount of charge transferred to surface [76]. The JCI 176 Charge Measuring Sample Support (connected with JCI 155 v5) provides a convenient unit to support film and layer materials (and also powder and liquids). The samples of conducting copolymer blends (LDPE/conducting copolymers) i.e. 45 x 54 mm^2 blown film were used for measurement of static decay time on John Chubb Instrument (Model JCI 155 v5) by measuring the time on applying the positive as well as negative high corona voltage of 5000 V on the surface of material to be tested and recorded the decay time at 10 % cutoff. A fast response electrostatic field meter observes the voltage received on the surface of sample and measurements were to observe how quickly the voltage falls as the charge is dissipated from the film. The basic arrangement for measuring the corona charge transferred to the test sample during corona charge decay measurements is shown in Figure 24. Charge is measured as a combination of two components-'conduction charge' and 'induction charge'. The 'conduction' component is that which couples directly to the sample mounting plates within the time of application of corona charging and the time for the plate carrying the corona discharge points to move away.

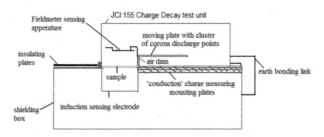

Figure 25. Schematic arrangement of JCI 155 v5 on JCI 176 charge measuring sample support.

The 'inducting' relates to the charge that has been deposited but has not coupled out directly to the mounting plates and the total charge transferred to the sample can be measured as:

$$Q_{total} = Q_{(conduction)} + f * Q_{(induction)} \qquad (6)$$

where the factor 'f' is actually close to 2.2. This factor can be determined experimentally. The film and layer polymeric samples are easily mounted in the JCI 176 between the two hinged flat metal plates. The aperture in the sample mounted plates, to which the conduction charge is measured, are 5 mm larger all round than the 45 x 54 mm^2 test aperture of the JCI 155. The JCI 155 Charge Decay Test unit sits on top of the JCI 176 Charge Measuring Sample Support into the recess between the boundary edges. The measurements are recorded in the form of graphs (ESD/JCI-graphs) which show the decay of surface voltage with respect to decay time.

12. Synthesis of poly(aniline-co-1-amino-2-naphthol-4-sulphonic acid)

Copolymers of 1-amino-2-naphthol-4-sulphonic acid (ANSA) and aniline of varying composition (i.e. by varying the co-monomer feed compositions in the initial feed) were synthesised by chemical oxidative polymerization in the presence of PTSA. Polymerization was initiated by the drop wise addition of ammonium persulphate solution (0.1 M APS in distilled water). The polymerization was carried out at a temperature of 0 C for a period of 4-6 h. Their copolymers were synthesised by varying the molar ratio of co-monomers in the initial feed. The synthesized copolymers were isolated from reaction mixture by filtration and washed with distilled water to remove oxidant and oligomers.

PTSA doped copolymers of aniline and ANSA (poly(AN-co-ANSA) in 80:20 molar ratio and 50:50 molar ratio is abbreviated as PANSA2-PTS and PANSA5-PTS respectively whereas PTSA doped polyaniline is abbreviated as PANI-PTS.

13. Preparation of LDPE-Conducting Copolymer Film

Composites of copolymers with LDPE were prepared by melt blending method. Required amount of LDPE and copolymers were loaded in internal mixer for 20-30 minutes at around 60 rpm. Blending of copolymers with LDPE was carried out in twin-screw extruder at the temperature range from 140-150°C by melt mixing method. The blown film of the copolymer/LDPE composite was made by Haake Blown Film instrument at the temperature range of 160°C where speed of screw was maintained at 40 rpm. PTSA doped copolymers of aniline and ANSA (poly(AN-co-ANSA) in 80:20 and 50:50 molar ratio blended with LDPE is abbreviated as PANSA2-PTS/LDPE and PANSA5-PTS/LDPE respectively.

14. Characterization

14.1. Characterization of PTSA doped PANI and copolymers of AN and ANSA

ANSA is a tri-functional monomer having three functional groups (i.e. -NH$_2$, -OH and –SO$_3$H) along with two fused benzene rings.

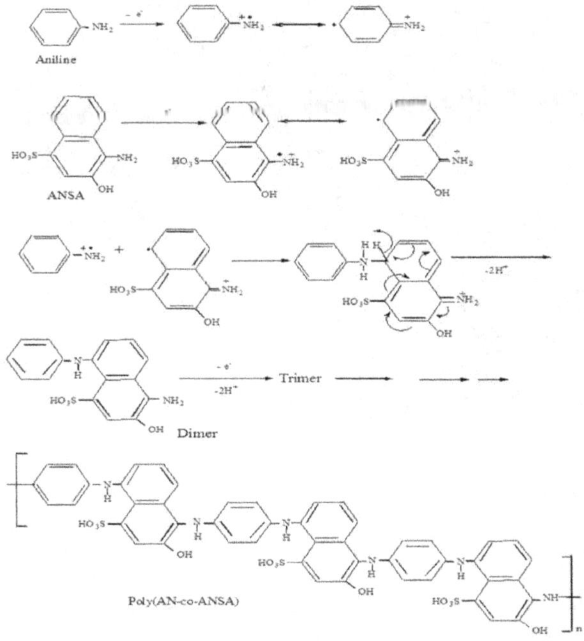

Figure 26. Proposed mechanism during the copolymerization of aniline and ANSA in the presence of p-toluene sulphonate (Reproduced with permission from Ref. 80, Copyright 2009 John Wiley & Sons).

This monomer can be copolymerized with aniline to give different materials and it has been observed that the participation of functional groups (-NH$_2$ and –OH) in the polymerization depends upon the reaction conditions. It is proposed that polymerization of ANSA in the presence of PTSA occurred selectively through –NH$_2$ group (figure 26) as confirmed by structural characterization (FTIR and NMR spectroscopy) [77].

14.1.1. Morphological Characterization

Figure 27 shows SEM micrographs of PTSA copolymers of ANSA and AN. PANI-PTS show a globular sponge like structure (Figure 27a) and morphology changed with varying copolymer composition. PTSA doped copolymers of aniline and ANSA exhibit hollow tube like morphology. The use of 1-amino-2-naphthol-4-sulphonic acid as a co-monomer as well as nature of external dopant played an important role for achieving the tubular morphology. In case of PTSA doped copolymers of aniline and ANSA in ratio of 80:20 (PANSA2-PTS), the globular morphology of the resultant copolymer tend to change to the tube forms (Figure 27b).

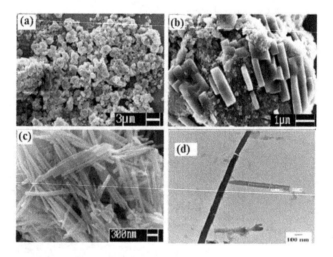

Figure 27. SEM image of (a) PANI-PTS (b) PANSA2-PTS; (c) PANSA5-PTS and (d) TEM image of PANSA5-PTS (Reproduced with permission from Ref. 77, Copyright 2009 John Wiley & Sons).

However, well defined tubes were formed when molar ratio of aniline/ANSA was 50:50 in the presence of PTSA (Figure 27c). The difference in the morphology between polyaniline and its copolymers with ANSA may be related to the different reactivities of the two monomers, nature of reaction media and reaction route.

TEM image of PTSA doped copolymer of aniline and ANSA in 50:50 molar ratios (Figure 26d) shows that these tubes are hollow with outer diameter of 80-90 nm.

14.1.2. Conductivity

Room temperature conductivity values of PTSA doped samples are summarised in Table 5, which reveals that the room temperature conductivity of PANI-PTS was found to be better than PTSA doped copolymers.

Sample Designation	Room temperature conductivity (S/cm)	Thermal stability (°C)
PANI-PTS	1.72	200
PANSA2-PTS	4.48×10^{-1}	195
PANSA5-PTS	1.98×10^{-2}	188

Table 5. Room temperature conductivity and thermal stability.

On increasing the molar ratio of ANSA in copolymer, conductivity tends to decrease accordingly due to the presence of three functional groups in ANSA unit which exerted a strong steric effect on the doping process hence induces additional deformation along the polymer backbone.

14.2. Characterization of LDPE/Conducting Polymer composite film

14.2.1. Thermal Properties

These copolymers (PANSA2-PTS/PANSA5-PTS) can be melt blended with conventional polymers like LDPE. Figure 28 shows the TG traces of blown films of PTSA doped copolymer/ LDPE composites. The degradation temperature of pure LDPE blown film was around 400°C. Thermal stability of the blown film of copolymer/LDPE blends (0.5-1.0 wt % loading) was also found to be same as LDPE.

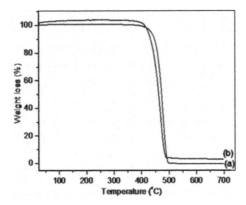

Figure 28. TG traces of (a) LDPE/PANSA2 -PTS and (b) LDPE/PANSA5-PTS films at 1.0 wt. % loading.

14.2.2. Mechanical properties

The mechanical properties of PTSA doped poly (AN-co-ANSA)/LDPE film was measured and the results are summarised in the Table 6. In case of pure LDPE film, the tensile modulus and yield stress were 141 MPa and 16.3 MPa respectively. However, the inclusion of poly (AN-co-ANSA) in LDPE led to decrease in both tensile modulus and yield stress depending upon the molar ratio of ANSA in the copolymer chain as well as type of dopant.

Sample Designation	Loading of copolymers In LDPE (wt. %)	Tensile modulus (MPa)	Yield Stress (MPa)	Ultimate elongation (%)	Conductivity of LDPE/ copolymers film (S/cm)
LDPE	1.0	141	16.3	187	"/10^{-12}
LDPE/PANSA2-PTS	1.0	129	12.1	176	1.28×10^{-6}
	0.5	134	13.1	180	2.22×10^{-9}
LDPE/PANSA5-PTS	1.0	120	10.3	166	8.18×10^{-7}
nanocomposites	0.5	131	12.3	171	4.13×10^{-9}

Table 6. Mechanical and electrical properties of LDPE films in the absence/presence of conducting copolymers.

In the case of film prepared by composites of LDPE/conducting copolymer (99/1 w/w or 99.5/0.5 w/w), tensile modulus, yield stress and % elongation decreased (Table 6). Tensile modulus decreased from 141 MPa (LDPE) to 120 MPa (LDPE/PANSA5-PTS) at a concentration of 1.0 % w/w. Yield stress also decreased for 16.3 MPa (LDPE) 10.3 MPa (LDPE/ PANSA5-PTS). Similarly, the ultimate elongation also decreased in the same manner.

At a loading of 0.5 % (w/w) of PTSA doped copolymers in LDPE, tensile modulus also decreased from 134 MPa (for PANSA2-PTS/LDPE) to 120 MPa (for PANSA5/LDPE). Similarly, 0.5 wt. % loading of PTSA doped copolymers with LDPE, yield stress reduced from 13.1 MPa in case of PANSA2-PTS/LDPE to 12.3 MPa for PANSA5-PTS/LDPE film and the ultimate elongation was also found to be 180 % and 171 % for PANSA2-PTS/LDPE and PAN-SA5-PTS/LDPE composite films respectively. Hence, the mechanical strength of PTSA doped poly(AN-co-ANSA)/LDPE blended films was found to be better in case 0.5 wt.% loading of copolymers than that of 1.0 wt. % loading. Moreover, it has also been observed that mechanical properties of PTSA doped copolymers-LDPE film were different from that of self doped copolymers-LDPE films. Mechanical strength of the poly(AN-co-ANSA)/LDPE composites decreased with increasing the molar ratio of ANSA in the copolymer (Table 6).

14.2.3. Electrical Properties

Room temperature conductivity values of PTSA doped copolymers/LDPE composite film are summarised in Table 6. The room temperature conductivity of copolymers of aniline with ANSA decreased from 4.48×10^{-1} to 1.98×10^{-2} S/cm depending on the molar ratio of

ANSA in the copolymer feed and type of dopant. The conductivity values copolymers were found to be 4.48×10^{-1} S/cm and 1.98×10^{-2} S/cm for PANSA2-PTS and PANSA5-PTS respectively (Table 5). On blending with LDPE at 1.0 wt %, conductivity value decreased from 1.28 $\times 10^{-6}$ S/cm to 8.18×10^{-7} S/cm respectively. When the loading level of copolymers with LDPE reduced from 1.0 wt % to 0.5 wt %, the conductivity of the resultant composites decreased. 0.5 % (w/w) loading of LDPE films based on PANSA2-PTS and PANSA5-PTS had conductivity value in the order of 2.22×10^{-9} S/cm and 4.13×10^{-9} S/cm respectively.

14.2.4. Morphological Characterization

Figure 29 show the SEM images of LDPE films in the presence of PTSA doped copolymers at 0.5 wt. % loading. When these copolymers were blended with LDPE, the copolymer domains were found to disperse in the LDPE matrix as evident by the appearance of tubes and needle like granules in the LDPE matrix (Figure 29). In addition, the formation of the conducting path is evident and agrees with the results relating to electrical conductivity of the composites. In copolymer composites (matrix and dispersed phase), the level of interaction between the two components and mode of dispersion in the matrix, influence the electrical and mechanical properties of the composites [79]. The SEM micrographs of the LDPE/copolymer film showed two different phases i.e. conducting copolymer and non conducting matrix (LPDE). Interconnection of conducting phase in the non-conducting matrix creates a conducting path along the LDPE matrix.

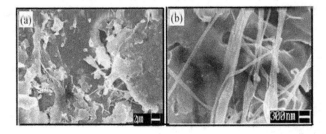

Figure 29. SEM images of (a) PANSA2-PTS/LDPE, (b) PANSA5-PTS/LDPE composite films at 0.5% (w/w) loading (Reproduced with permission from Ref. 77, Copyright 2009 John Wiley & Sons).

Moreover, it has also been observed that the conduction mechanism and transportation of charge carrier in the blends depend on the loading level and mode of dispersion of the conducting materials. PANSA5-PTS nanotubes at 0.5 wt. % loading with LDPE, the surface conductivity was found to be in the order of 10^{-9} S/cm, which is suitable for their use in ESD protection applications.

Hence, it may be presumed that when the sufficient amount of conducting material is loaded in the polymer matrix, the conducting particles get closer and form linkage which makes an easy path for conduction of charge carrier throughout the blend which shows sufficient loading and good dispersion of conducting material in the polymer matrix (i.e. LDPE). While in the case of very low loading of conducting material in the polymer matrix, the gap between conducting particles in the polymer matrix is large with the result that no conduction path in the blend.

Hence, the conductivity of films based on blends depends on the morphology of conducting material. The nanotubular or fibre like morphology of conducting materials which form a network in the whole blend facilitate the conduction of charge carrier through the continuous structure of the chain of conducting material in the insulating matrix at very low loading of conducting material in the insulating matrix.

15. Antistatic Behaviour of LDPE/Copolymer Film

The results of static decay time on application of positive/negative voltage of 5000 V and recording the decay time on going down to 500 V are summarised in Table 7. It was observed that blank LDPE film shows a static decay time of 120.1 sec. It decreased upon addition of copolymer and was found to be dependent on the amount of copolymer. LDPE film having 1.0 % (w/w) and 0.5 % (w/w) of PANSA2-PTS showed a decay time of 0.1 sec. and 1.4 sec. respectively at 10 % cut-off. However, the PANSA5-PTS/LDPE film showed a static decay time of 0.8 sec. at a loading of 0.5 wt. % and 0.2 sec. at 1.0 wt.% loading [80]. Any material which showed a static decay time less than 2.0 sec passes the criteria for its use as antistatic material. Based on the above observations, we can say that LDPE film prepared by blending of conducting copolymer based on AN and ANSA at 1.0 % w/w loading, can be used as an effective antistatic film. Similar measurements were recorded with copolymer composite film with a cut-off value of 50 % and the results are summarised in Table 7.

Sample Designation	Loading of copolymers in LDPE (wt. %)	Static decay time (at 10 % Cut off) (Sec.)		Static decay time (at 50 % Cut off) (Sec.)	
		Positive voltage	Negative voltage	Positive voltage	Negative voltage
Blank LDPE	--	120.1	110.8	94.9	93.1
LDPE/PANSA2-PTS	1.0	0.1	0.1	0.01	0.01
	0.5	1.4	1.5	0.2	0.3
LDPE/PANSA5-PTS nanocomposites	1.0	0.2	0.1	0.01	0.01
	0.5	0.8	0.9	0.1	0.1

Table 7. Antistatic behaviour of LDPE/copolymer composite films.

Static decay measurements were also performed on John Chubb Instrument (JCI 155 v5) charge decay test unit by measuring the time on applying the positive as well as negative high corona voltage of 5000 V on the surface of material to be tested and recorded the decay time at 10 % cut off. A fast response electrostatic field meter observes the voltage received on the surface of sample and measurements were to observe how quickly the voltage falls as the charge is dissipated from the film. Graphs obtained from these experiments have been shown in the Figure 30, which show the decay of surface voltage and decay time.

The surface voltage and surface charge received by the materials depend on nature of materials. When positive or negative high corona voltage (i.e. 5000 V) was applied to the surface of the material, only a limited amount of voltage was received by the blend depending on the nature of materials. When high corona voltage was applied on the surface of insulating material, only some voltage was drained away and greater amount of voltage were retained on its surface. This surface voltage decays at particular time. Moreover, the surface charge received by the blends was also calculated during the experiment.

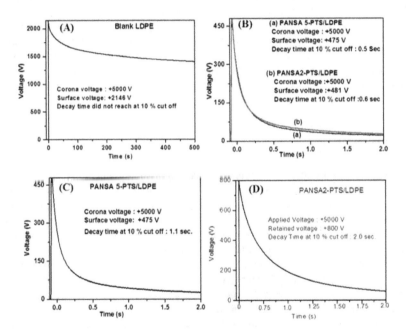

Figure 30. ESD-graphs of LDPE film in the presence of AN-ANSA copolymers (A) blank LDPE film, (B) Copolymer/LDPE nanocomposites film at 1.0 wt.% loading, (C) PANSA5-PTS at 0.5 wt. % loading and (D) PANSA2-PTS at 0.5 wt. % loading (Reproduced with permission from Ref. 77, Copyright 2009 John Wiley & Sons).

Hence the charge retention capability of conducting materials was found to be very low thus they quickly dissipate this surface charge. The static decay time of blank LDPE film was found to be very high on applying the positive and negative corona voltage of 5000 V. The peak at 2146 V indicate that the LDPE film has received 2146 V at the surface corresponding to 55.79 nC of static charge, which get dissipated very slowly and was not found to be able to dissipate it up to 10 % cut off as shown in the Figure 30A. Due to insulating nature of the material, lot of charges were found to be retained on the surface of LDPE film. Blending of 1.0 wt. % of conducting copolymer with LDPE, decreases the charge retention capability by reducing the decay time. In case of PANSA5-PTS/ LDPE film at 1.0 wt. % loading of PANSA 5-PTS in LDPE, the peak started at 475 V on applying the voltage of +5000 V, which indicates that the voltage received at the surface is only 475 V corresponding to 9.51 nC of charges which dissipated quickly, around 0.5 sec., at 10% cut off as shown in the Figure 30B (curve a).

Similar trend has been found for negative polarity charging at the same corona voltage. In the case of film samples prepared by blending of 0.5 wt. % PANSA5-PTS with LDPE, 475 V of voltage and 10.45 nC of charges were received by its surface which was dissipated up to 10 % cut off in 1.1 sec (Figure 30C). on the other hand, LDPE + 0.5 wt. % PANSA2-PTS showed + 800 V of surface voltage received by the composite on applying the voltage of +5000 V which showed the large decay time (2.0 sec at 10 % cut off). Similar behaviour was observed at negative polarity charging. Hence the ESD protection performance of the conducting blends not only depends on the loading level of conducting materials but also depend on the morphology and dispersion of conducting materials in the polymer matrix. Nanocomposites based on LDPE/PANSA5-PTS film showed better ESD performance as compared to LDPE/PANSA2-PTS film.

16. Conclusions

PANI/SiO$_2$ nanocomposites were prepared by chemical oxidation polymerization of aniline and SiO$_2$ by using ammonium persulfate (APS) as an oxidant in the presence of phosphoric acid/PFOA medium. FTIR, UV-Visible, cyclic voltammetry and SEM techniques confirmed the interaction of PANI with SiO$_2$ particles. The excellent corrosion protection performance by PSC coated mild steel could be due to the strong adherence of polymer film which uniformly covers the entire electrode surface as has shown by the surface morphology. The corrosion current densities were lowered several orders of magnitude with these coatings. The coating had good protective efficiency which increased with increasing the loading of PSC to the maximum of 99 % at 6.0 wt.% loading and reduced to about 89.93 % after 60 days of immersion in highly corrosive environment confirming the improved coating performance. Weight loss method also revealed that PSC coated samples showed very low weight loss as well as negligible corrosion rate as compared to PANI coated samples at same immersion time, which indicates the better protection and adhesion of PSC onto the mild steel surface as compared to PANI in strong acidic condition.

In order to improve anticorrosion performance of iron in 3.5 % NaCl aqueous medium, preparation of highly hydrophobic polyaniline-SiO$_2$ nano-composites (HPSC) have also been carried out by chemical oxidation polymerization. Water repellent property of the PSC has been developed by using fluorinated dopant i.e. perfluoro-octanoic acid (PFOA). HPSC coating were evaluated for protection of mild steel from corrosion in 3.5 % NaCl aqueous solution. Suitable coating with HPSC was formed on mild steel using epoxy resin by powder coating technique which showed the contact angle in the range of 115°. Corrosion protection efficiency of mild steel coated HPSC in 3.5 % NaCl aqueous solution has been evaluated using Tafel Extrapolation method, surface morphology, salt spray test and weight loss methods. The results reveals that the HPSC coating showed the significant reduction in the corrosion current density reflects the better protection of mild steel in marine environment. The coating had good protective efficiency which increased with increasing the loading of HPSC to the maximum of 96 % at 6.0 wt.% loading and reduced to about 93.3 % after 60 days of immersion in 3.5 % NaCl solution confirming the improved coating performance.

Presence of SiO_2 nanoparticles entrapped in PANI chain which was evident my surface morphology of composite coating, provide the reinforcement to PANI chain which reduce the degradation of polymer chain in corrosive environment. PSC/HPSC coating protect metal by dual mechanism by forming passivating layer as well as act as a physical barrier. Furthermore the role of powder coating technique for achieving high quality, durable and good anticorrosive coatings have also been explained. These studies revealed that the polyaniline-SiO_2 nanocomposites has excellent corrosion protection properties and it can be considered as a potential material for corrosion protection of mild steel in corrosive medium like.1.0 M HCl as well as 3.5 % NaCl solution.

In order to carry out the effective use of conducting polymer for antistatic application, nanocomposites based on poly(aniline -co- 1-amino-2-naphthol-4-sulphonic acid) (PANSA5-PTS) with low density polyethylene (LDPE) have been developed. The copolymer nanotubes of aniline and ANSA were synthesised in tosyl medium in 50: 50 molar ratio. Formation of nanotubes of copolymers was confirmed by morphological characterization using SEM and TEM. Dimension of nanotubes of PANSA5-PTS was found to be 80-90 nm. Blending of copolymers with LDPE was carried out in twin screw extruder by melt blending method by loading 0.5 wt. % and 1.0 wt. % of the conducting copolymer in LDPE matrix. The conductivity of the blown film of poly (AN-co-ANSA) /LDPE composites was found to be in the range of 1.28 x10^{-6} to 4.13x 10^{-9} S/cm. Thermo gravimetric traces of copolymers reveals that these copolymers were thermally stable from 180°C to 195°C. Such copolymers were successfully melt blended with LDPE and conducting film was prepared using film blending technique. Antistatic performance of PANSA5-PTS/LDPE nanocomposite have compared with PANSA2-PTS/LDPE composites to show the influence of nanotubes in composites. Static charge measurements carried out on the films shows that no charge is present on the surface. Copolymer/LDPE composites films (1.0 % w/w) showed static decay time in the order of 0.1 to 0.2 sec. at 10 % cut-off on recording the decay time from 5000 V to 500 V. When the loading level of copolymers in LDPE was reduced to 0.5 wt. %, only the nanocomposites based on PANSA5-PTS showed better good performance to ESD protection. Better antistatic behavior shown by these copolymers at very low loading in LDPE was investigated by their nanotubular morphology. Blending of 0.5 and 1.0 wt. % of PTSA doped copolymers with selective composition of ANSA and aniline with LDPE has a great potential to be used as effective antistatic films. The loading level, morphology of the conducting material, and its proper dispersion with insulating matrix affect the properties like surface conductivity, mechanical properties, and its performance to application for electrostatic charge dissipation.

Author details

Hema Bhandari, S. Anoop Kumar and S. K. Dhawan

*Address all correspondence to: skdhawan@mail.nplindia.ernet.in

CSIR–National Physical Laboratory, India

References

[1] Zhua, H., Zhonga, L., Xiaoa, S., & Gan, F. (2004). *Electrochim. Acta*, 49, 5161.

[2] Jones, D. A. (1992). *Principles and Prevention of Corrosion, Macmillan Publishing, Chap. 1*, New York.

[3] Lacroix, J. C., Camalet, J. L., Aeiyach, S., Chane-Ching, K. I., Petitjean, J., Chauveau, E., & Lacaze, P. C. (2000). *J. Electroanal. Chem.*, 481, 76.

[4] Kinlen, P. J., Menon, V., & Ding, Y. (1999). *J. Electrochem. Soc.*, 146, 3690.

[5] Kinlen, P. J., Ding, Y., & Silverman, D. C. (2002). *Corrosion*, 58, 490.

[6] de Souza, S., da Silva, J. E. P., de Torrosi, S. I. C., Temperani, M. L. A., & Torresi, R. M. (2001). *Electrochem. Solid State Lett.*, 4, B27.

[7] Samui, A. B., Patankar, A. S., Rangarajan, J., & Deb, P. C. (2003). *Prog. Org. Coat.*, 47.

[8] Dominis, A. J., Spinks, G. M., & Wallace, G. G. (2003). *Prog. Org. Coat.*, 48, 43.

[9] Sathiyanarayanan, S., Muthukrishnan, S., & Venkatachari, G. (2006). *Prog. Org. Coat.*, 55, 5.

[10] Plesu, N., Ilia, G., Pascariu, A., & Vlase, G. (2006). *Synth. Met.*, 156, 230.

[11] Su, S. J., & Kuramuto, N. (2001). *Synth. Met.*, 114, 147.

[12] Gurunathan, K., Amalnerker, D. P., & Trivedi, D. C. (2003). *Mater. Lett.*, 57, 1642.

[13] Sathiyanarayanan, S., Muthukrishnan, S., Venkatachari, G., & Trivedi, D. C. (2005). *Prog. Org. Coat.*, 53, 297.

[14] Deberry, D. W. (1985). *J. Electrochem. Soc.*, 132, 1027.

[15] Wessling, B. (1991). *Synth. Met*, 41, 907.

[16] Elsenbaumer, R. L., Lu, W. K., & Wessling, B. (1994). Seoul, Korea. *Int. Conf. Synth. Met.*, Abstract No. APL(POL)1.

[17] Wrobleski, D. A., Benicewicz, B. C., Thompson, K. G., & Byran, C. (1994). *J. Polym. Prepr. (Am. Chem. Soc., Div. Polym. Chem.)*, 35, 265.

[18] Spinks, G. M., Dominis, A. J., Wallace, G. G., & Tallman, D. E. (2002). *J. Solid State Electrochem*, 6, 85.

[19] Sathiyanarayanan, S., Azim, S. S., & Venkatachari, G. (2007). *Electrochimica Acta*, 52, 2068.

[20] Majumdar, G., Goswami, M., Sarma, T. K., Paul, A., & Chattopadhyay, A. (2005). *Langmuir*, 21, 1663.

[21] Chowdhury, D., Paul, A., & Chattopadhyay, A. (2005). *Langmuir*, 21, 4123.

[22] Feng, X. M., Yang, G., Xu, Q., Hou, W. H., & Zhu, J. J. (2006). *Macromol Rapid Commun*, 27, 31.

[23] Hasan, M., Zhou, Y., Mahfuz, S., & Jeelani, S. (2006). *Materials Science and Engineering: A*, 429, 181.

[24] Li, X., Dai, N., Wang, G., & Song, X. (2008). *J Appl Polym Sci*, 107, 403.

[25] Xia, H. S., & Wang, Q. (2003). *J Appl Polym Sci*, 87, 1811.

[26] Zengina, H., & Erkan, B. (2010). *Polym. Adv. Technol.*, 21, 216.

[27] Stejskal, J., Kratochvi'l, P., Armes, S. P., Lascelles, S. F., Riede, A., Helmstedt, M., Prokes, J., & Krivka, I. (1996). *Macromolecules*, 29, 6814.

[28] Al-Dulaimi, A. A., Hashim, S., & Khan, M. I. (2011). *Sains Malaysiana*, 40, 757.

[29] Beck, F., Michaelis, R., Scholoten, F., & Zinger, B. (1994). *Electrochimica Acta*, 39, 229.

[30] Camalet, J. L., Lacroix, J. C., Aeiyach, S., Chane-Ching, K., & Lacaze, P. C. (1998). *Synth. Met.*, 93, 133.

[31] Kilmartin, P. A., Trier, L., & Wright, G. A. (2002). *Synthetic Metals*, 131, 99.

[32] Meneguzzi, A. A. P., Ferreira, C. A., Pham, M. C., Delamar, M., & Lacaze, P. C. (1999). *Electrochim. Acta*, 44, 2149.

[33] Bhandari, H., Choudhary, V., & Dhawan, S. K. (2010). *Thin Solid Film*, 519, 1031.

[34] Kinlen, P. J., Menon, V., & Ding, Y. J. (1999). *Electrochem. Soc.*, 146, 3690.

[35] Talo, A., Passiniemi, P., Forse'n, O., & Yla''saari, S. (1997). *Synth. Met.*, 85, 1333.

[36] Wessling, B., & Posdorfer, J. (1999). *Electrochim. Acta.*, 44, 2139.

[37] Iribarren, J. I., Armelin, E., Liesa, F., Casanovas, J., & Aleman, C. (2006). *Material and corrosion*, 57, 683.

[38] Mc Andrew, T. P., Miller, S. A., Gilleinski, A. G., & Robeson, L. M. (1996). *Polym. Mater. Sci. Eng.*, 74, 204.

[39] Mc Hale, G., Shirtcliffe, N. J., Aqil, S., Perry, C. C., & Newton, M. I. (2004). *Phys. Rev.Lett.*, 93, 36102.

[40] Shirtcliffe, N. J., Mc Hale, G., Newton, M. L., Chabrol, G., & Perry, C. C. (2004). *Adv.Mater.*, 16, 1929.

[41] Wu, X. F., & Shi, G. Q. (2006). *J. Phys. Chem. B.*, 110, 11247.

[42] Jiang, L., Zhao, Y., & Zhai, J. (2004). *Angew. Chem. Int. Ed.*, 43, 4338.

[43] Jiang, W. H., Wang, G. J., He, Y. N., An, Y. L., Wang, X. G., Song, Y. L., & Jiang, L. (2005). Chem. J. Chin. Univ. (Chinese) , , 26, 1360.

[44] Han, J. T., Zheng, Y., Cho, J. H., Xu, X., & Cho, K. J. (2005). *Phys. Chem. B.*, 109, 20773.

[45] Soeno, T., Inokuchi, K., & Shiratori, S. (2004). *Appl. Surf. Sci.*, 237, 543.

[46] Li, H., Wang, X., Song, Y., Liu, Y., Li, Q. L., & Zhu, D. (2001). *Angew. Chem.*, 113, 1793.

[47] Wu, Y., Sugimura, H., Inoue, Y., & Takai, O. (2002). *Chem. Vap. Deposition*, 8, 47.

[48] Wang, S. T., Feng, L., & Jiang, L. (2006). *Adv. Mater.*, 18, 767.

[49] Qu, M. N., Zhang, B. W., Song, S. Y., Chen, L., Zhang, J. Y., & Cao, X. P. (2007). *Adv. Funct. Mater.*, 17, 593.

[50] Pan, Q. M., Jin, H. Z., & Wang, H. B. (2007). *Nanotechnology*, 18, 355605.

[51] Yabu, H., Takebayashi, M., Tanake, M., & Shimomura, M. (2005). *Langmuir*, 21, 3235.

[52] Li, X., Reindhout, D., & Crego-Calama, M. (2007). *Chem. Soc. Rev*, 36, 1350.

[53] Callies, M., & Quere, D. (2005). *Soft Matter*, 1, 55.

[54] Ma, M. L., & Hille, R. M. (2006). *Curr. Opin. Colloid Interface Sci.*, 11, 193.

[55] Blossey, R. (2003). *Nat. Mater.*, 2, 301.

[56] Anoop, Kumar. S., Bhandari, H., Sharma, C., Khatoon, F., & Dhawan, S. K. (2012). *Polymer International.*

[57] Anoop, Kumar. S., Bhandari, H., Sharma, C., Khatoon, F., & Dhawan, S. K. (2012). *Polymer (Paper under process).*

[58] Stern, M., & Geary, A. (1957). *J. Electrochem Soc.*, 104, 56.

[59] Spinks, G. M., Dominis, A. J., Wallace, G. G., & Tallman, D. E. (2002). *J. Solid State Electrochem.*, 6, 85.

[60] Silverstein, R. M., & Webster, F. X. (2002). *Spectrometric identification of organic compounds* (sixth edition), John Wiley and Son, Wiley, India, 165.

[61] Kim, H., Foster, C., Chiang, J., & Heegar, A. J. (1989). *Synth. Metals.*, 29, 285.

[62] Pron, A., & Rannou, P. (2002). Progr. Polym. Sci. ; , 27, 135.

[63] Huerta-Vilca, D., Moraes, S. R., & Motheo, A. J. (2003). *J. Braz. Chem. Soc.*, 14, 52.

[64] Mc Andrew, T. P., Miller, S. A., Gilleinski, A. G., & Robeson, L. M. (1996). *Polym. Mater. Sci. Eng.*, 74, 204.

[65] Kusy, R. P. (1986). in Metal filled polymers, edited by S.K. Battacharya (Dekker New York , 1.

[66] Rosner, R. B. (2001). *IEEE Transaction on device and material reliability*, 1.

[67] Edenbaum, J., & Reinhold, V. N. *Plastics Additives and Modifiers Handbook*, New York, 615.

[68] Narkis, M., Lidor, G., Vaxman, A., & Zuri, L. (1999). *J. Electrostatics*, 47, 201.

[69] Li, C., Liang, T., Lu, W., Tang, C., Hu, X., Cao, M., & Liang, J. (2004). *Composite Sci. Tech.*, 64, 2089.

[70] Narkis, M., Ram, A., & Stein, A. Z. (1980). *J. Appl. Polym. Sci.*, 25, 1515.

[71] Kobayashi, T., Wood, A., Takemura, A., & Ono, A. (2006). *Barbara Jnl. Electrostat.*, 64, 377.

[72] Yukishige, H., Koshima, Y., Tanisho, H., & Kohara, T. (1996). *US patent 5571859*.

[73] Angelopoulos, M. (2001). Conducting polymers in microelectronics. *IBM Journal of Research and Development*, 45(1), 57-76.

[74] Laska, J., Żak, K., & Prón, A. (1997). Conducting blends of polyaniline with conventional polymers. *Synthetic Metals*, 84(1), 117-118.

[75] Mitzakoff, S., & De Paoli, M. A. (1999). European Polymer Journal ., 35(10), 1791-1798.

[76] Chubb, J. N. (2002). *J. Electrostatics*, 54, 233.

[77] Bhandari, H., Bansal, V., Choudhary, V., & Dhawan, S. K. (2009). *Polymer International*, 58, 489.

[78] Chen, S. A., & Hwang, G. W. (1996). *Macromolecules*, 29, 3950.

[79] Bhandari, H. (2011). Synthesis, Characterization and Evaluation of Conducting Copolymers for Corrosion Inhibition and Antistatic Applications. *PhD thesis*, Indian Institute of Technology, Delhi, India.

[80] Bhandari, H., Singh, S., Choudhary, V., & Dhawan, S. K. (2011). *Polymers for Advanced Technologies*, 22(9), 1319-1328.

[81] Strumpler, R., & Glatz-Reichenbach, J. (1999). Conducting polymer composites. J. Electroceramics ., 3(4), 329-346.

Chapter 3

Electroconductive Nanocomposite Scaffolds: A New Strategy Into Tissue Engineering and Regenerative Medicine

Masoud Mozafari, Mehrnoush Mehraien,
Daryoosh Vashaee and Lobat Tayebi

Additional information is available at the end of the chapter

1. Introduction

Nanocomposites are a combination of a matrix and a filler, where at least one dimension of the system is on the nanoscale being less than or equal to 100 nm. Much work has focused on the construction of nanocomposites due to the structural enhancements in physico-chemical properties, and functionality for any given system [1-6]. The physico-chemical enhancements result from the interaction between the elements being near the molecular scale. Nanocomposite materials have also received interest for tissue engineering scaffolds by being able to replicate the extracellular matrix found *in vivo*. Currently, researchers have created composite materials for scaffold formation which incorporate two or more materials. Some of these materials consist of minerals for bone tissue engineering including calcium, hydroxyapatite, phosphate, or combinations of different polymers, such as poly (lactic acid), poly (ε-caprolactone), collagen and chitosan, and many other different combinations [7-9]. Other work has focused on doping the polymer scaffolds with specific growth hormones or adhesion sequences to influence how cells attach to the scaffold and cause the scaffold to become a drug delivery vehicle for different kind of tissue engineering applications [10]. Among different materials used in preparation of nanocomposits, conducting polymers are one of the effective materials that can be employed to facilitate communication with neural system for regenerative purposes.

However, the major obstacle concerning the electrically conducting polymers has been the difficulty associated with the processing of them [11]. To overcome this problem, most researchers have electrospun conducting polymers by blending them with other spinnable polymers, compromising the conductivity of the nanocomposite fibres [12-16]. Blending of conducting polymers with other polymers positively affects the properties of the resultant nanocomposite fibres. In addition, sometimes for making benefid from condicting polymers and the specific properties of them we can have just a small thib coating of the polymer on the surface of nanocomposite.

The term of "Tissue Inducible Biomaterials" has been recently applied based on the principles of biology and engineering to design nanocomposite scaffolds that restore, maintain or improve the general function of damaged tissues. To gain tissue induction activity and assist tissue regeneration, the nanocomposite scaffolds need to be designed based on nanostructural properties, surface modifications or incorporation of molecules into them. Among different approaches and materials for the preparation of scaffolds, get benefit from conducting polymers seems to be more interesting and promising. Electroconductive polymers exhibit excellent electrical properties and have been explored in the past few decades for a number of applications. In particular, due to the ease of synthesis, cytocompatibility, and good conductivity, some kind of conducting polymers have been extensively studied for biological and medical applications. Different forms of conducting polymers such as polypyrrole (PPy), polythiophene (PT), polyaniline (PANI), poly(3,4-ethylenedioxythiophene) (PEDOT) etc. are used in our daily life due to their uniqe properties which can be applied in different applications. These materials have a conjugated π electron system with "metal-like" electrical conductivity. Due to the rich chemistry of conducting polymers, they have attracted the attention of many researchers and leading to the publication of thousands of papers. The most important property of conducting polymers is their electrical conductivity, so the first approach is to study their electrical-related biological behaviors. Neurons are well known for the membrane-potential-wave style signal transduction. Hence, early studies were focused on the electrical stimulation to the neuron cells using conducting polymers as electrodes. The results showed that the electric conducting polymers can be used as biological electrodes and the neuron growth can be enhanced under an electrical field.

Using conducting polymers in nanocomposite scaffold design is relatively new in tissue engineering applications [17]. It has been demonstrated that these conducting nanocomposites are able to accept and modulate the growth of different cell types [18] including endothelial cells, [19] nerve cells [20] and chromaffin cells [21]. It has been demonstrated that using conducting nanocomposite scaffolds are most promising in nerve tissue engineering. These electroconductive polymers have been recognized as potential nanocomposite scaffold materials to electrically stimulate tissues for therapeutic purposes in tissue engineering scaffolds. Based on the literature search within the last decade, the present chapter summarized the strategy of electroconductive nanocomposite scaffolds for tissue engineering and regenerative medicine purposes.

2. Conductive polymers

2.1. General approaches and considerations

Conducting polymers are a special class of materials with electronic and ionic conductivity [22]. The structures of the widely used conducting polymers are depicted in Fig. 1 [23]. These polymers have immense applications in the fields of drug delivery, neuroprosthetic devices, cardiovascular applications, bioactuators, biosensors, the food industry and etc.

One of the first electrically conducting polymers, polypyrrole (PPy) was introduced in the 1960s, but little was understood about this polymer at that time [24]. In 1977, a research team reported a 10 million-fold increase in the conductivity of polyacetylene doped with io-dine as the first inherently conducting polymer [25,26]. Unlike polyacetylene, polypheny-lenes, are known to be thermally stable as a result of their aromaticity [27]. Polyheterocycles, such as PPy, polythiophene (PT), polyaniline (PANI), and poly(3,4-ethylenedioxythiophene) (PEDOT), developed in the 1980s, have since emerged as another class of aromatic conduct-ing polymers that exhibit good stabilities, conductivities, and ease of synthesis [28]. Table 1 shows a list of different conducting polymers and their conductivities [29].

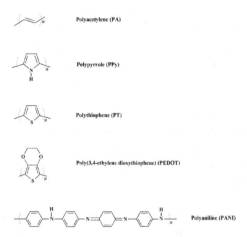

Figure 1. Chemical structures of various conducting polymers

Conducting polymers have an inherently unstable backbone, resulting from the formation of alternate single and double bonds along with the monomer units during polymerization. The delocalized π bonding electrons, produced across the conjugated backbone, provide an electrical pathway for mobile charge carriers which are introduced through doping. Conse-quently, the electronic properties, as well as many other physicochemical properties, are de-termined by the structure of the polymer backbone and the nature and the concentration of the dopant ion [30].

Conducting polymer	Maximum Conductivity (Siemens/cm)	Type of doping
Polyacetlene (PA)	200-1000	n,p
Polyparaphenylene (PPP)	500	n,p
Polyparaphenylene sulfide (PPS)	3-300	p
Polyparavinylene (PPv)	1-1000	p
Polypyrrole (PPy)	40-200	p
Polythiophene (PT)	10-100	p
Polyisothionaphthene (PITN)	1-50	p
Polyaniline (PANI)	5	n,p

Table 1. Some of the common conducting polymers and their conductivity [29].

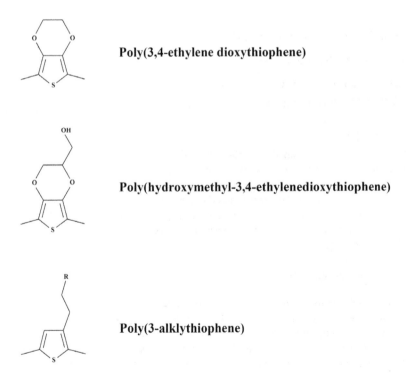

Poly(3,4-ethylene dioxythiophene)

Poly(hydroxymethyl-3,4-ethylenedioxythiophene)

Poly(3-alklythiophene)

Figure 2. Typical monomer structures used to fabricate Poly(3,4-ethylene dioxythiophene), Poly(hydroxymethyl- 3,4-ethylenedioxythiophene) and Poly(3-alklythiophene) [30]

Conjugated aliphatics, including polyacetylene, and benzene derivatives such as PANI, have been largely ruled out for biomedical applications due to their oxidative degradation in air and the cytotoxic nature of their by-products. Although recent research has shown that the emeraldine salt of PANI (EPANI) can be successfully fabricated in a biocompatible form [31,32], modern biomedical conducting polymers are typically composed of heterocyclic aromatics, such as derivatives of thiophene and pyrrole [33,34]. Specifically, PEDOT and PPy have been widely studied for their superior environmental and electrochemical stability [35-37]. Fig. 2 shows the chemical structure of various thiophene derivatives including EDOT, EDOT-MeOH and 3-alkylthiophene [30].

2.2. Surface modification of conducting polymers

For biomedical approaches, sometimes we need to modify the outer surface of the materials to induce special features. Conducting polymers can be also modified to enhance the functionality of nanocomposites. The surface modifications of conducting polymers have some concerns including:

- Enhancement of the charge transport of carriers between the implant and tissue

- Mediating the large difference in mechanical modulus

- Improvement of biodegradation

- Decreasing the impedance to enhance the sensitivity of the recording site

- Cell response enhancement

- Bioactivity enhancement

Surface modification and functionalization of conducting polymers with different biomolecules or dopants has allowed us to modify them with biological sensing elements, and to turn on and off different signalling pathways required for cellular processes. In this way, conducting polymers can show significant enhancement in cell proliferation and differentiation. Thus, conducting polymers provides an excellent opportunity for fabrication of highly selective, biocompatible, specific and stable nanocomposite scaffolds for tissue engineering of different organs [39,40].

2.3. General use of conducting polymers

A range of applications for conducting polymers are currently being considered, such as the development of tissue-engineered organs [41], controlled drug release [42], repair of nerve chanels [43], and the stimulation of nerve regeneration [44]. In addition, electrically active tissues (such as brain, heart and skeletal muscle) provide opportunities to couple electronic devices and computers with human or animal tissues to create therapeutic body–machine interfaces [45]. The conducting and semiconducting properties of this class of polymers make them important for a wide range of applications. The important properties of various conducting polymers and their potential applications are discussed in Table 2 [23].

Conducting polymer	properties	applications
polypyrrole (PPy)	Highly conductive	biosensors
	Opaque	drug delivery
	Brittle	bioactuators
	Amorphous structure	Nerve tissue engineering
		Cardiac tissue engineering
		Bone tissue engineering
polythiophenes (PT)	Good electrical conductivity	Biosensors
	Good optical property	Food industry
polyaniline (PANI)	A semiflexible rod polymer	Biosensors
	Requires simple doping/dedoping chemistry	Drug delivery
		Bioactuators
	Exists as bulk films or dispersions	Nerve tissue engineering
	High conductivity up to 100 S/cm	Cardiac tissue engineering
poly(3,4-ethylenedioxythiophene) (PEDOT)	High temperature stability	Biosensors
	Transparent conductor	Antioxidants
	Moderate band gap	Drug delivery
	Low redoxpotential	neural prosthetics
	conductivity up to 210 S/cm	

Table 2. Properties and applications of some common conducting polymer [23]

2.4. Conductivity mechanism

Generally, polymers with loosely held electrons in their backbones can be called conducting polymers. Each atom on the backbone has connection with a π bond, which is much weaker than the σ bonds in the backbone. These atoms have allways a conjugated backbone with a high degree of π-orbital overlap [46]. It is known that the neutral polymer chain can be oxidized or reduced to become either positively or negatively charged through doping process [47]. It is also known that conducting polymers could not be perfectly conductive without using dopants, and doping of π-conjugated polymers results in high conductivity [24]. The doping process is influenced by different factors such as polaron length, chain length, charge transfer to adjacent molecules and conjugation length [46]. There have been different dopants for the addition of H⁺ (protonation) to the polymers. For example, strong inorganic hydrochloric acid (HCl), organic and aromatic acids containing different aromatic substitution have been used as dopants for PANI. It is also reported that the surface energies of the doped conducting polymers vary greatly, depending on the choice of the dopants and doping level. Recently, PPy doped with nonbiologically active dopants (tosylate) and it has been characterized for biological interactions as they can trigger cellular responses in biological applications. However, the incorporation of more biologically active dopants can significantly modify PPy-based nanocomposites for biomedical applications [48].

One of the most important challenges of nanocomposite scaffolds based on conducting polymers is their inherent inability to degrade in the body, which may induce chronic inflammation [49]. Hence, belending of conducting polymers with biodegradable polymers seems to solve the problem. PPy and PANI are the most importan conducting polymers for tissue engineering, and they are important in terms of their biocompatibility and cell signaling especially for nerve tissue engineering [24].

2.5. Polypyrrole

PPy is among one of the first conducting polymers that studied a lot for its effect on the behaviour of cells. This material has been reported to support cell adhesion and growth of different cells [50]. This conducting synthetic polymer has numerous applications in tissue engineering and drug delivery. Recently, Moroder *et al.* [51] studied the properties of polycaprolactone fumarate–polypyrrole (PCLF–PPy) nanocomposite scaffolds under physiological conditions for application as conductive nerve conduits. In their study, PC12 cells cultured on PCLF–PPy nanocomposite scaffolds were stimulated with regimens of 10 μA of either a constant or a 20 Hz frequency current passed through the scaffolds for 1 h per day. The surface resistivity of the scaffolds was 2 kΩ and the nanocomposite scaffolds were electrically stable during the application of electrical stimulation. As can be seen in Fig. 3, *in vitro* studies showed significant increases in the percentage of neurite bearing cells, number of neurites per cell and neurite length in the presence of electrical stimulation compared with no electrical stimulation. They concluded that the electrically conductive PCLF–PPy nanocomposite scaffolds possed the material properties necessary for application in nerve tissue engineering.

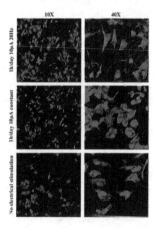

Figure 3. Fluorescence microscopy of PC12 cells at 10x and 40x magnification after undergoing different electrical stimuli treatment regimens for 48h

2.6. Polyaniline

PANI is an oxidative polymeric product of aniline under acidic conditions and is commonly known as aniline black [52]. The exploration of PANI for tissue-engineering applications has progressed more slowly than the development of PPy for similar applications. However, recently there has been more evidence of the ability of PANI and PANI variants to support cell growth [53]. Recently, Fryczkowski *et al.* [54] synthesized three-dimensional nanocomposite fibres of poly(3-hydroxybutyric acid) (PHB) and dodecylbenzene sulfonic acid (DBSA) doped polyaniline in chlorophorm/trifluoroethanol mixture, using electrospinning method. The morphology, electro-active properties and supermolecular structure of nanofibres webs have been analyzed and discussed. Obtained nanofibres are potentially applicable as nanocomposite scaffolds for tissue engineering. According to their results, there were limitations in composition of blended system and the PHB:PANI:solvent ratio needed to be optimized in order to obtain reasonable spinnability of compositions, and even small amount of PANI caused changes in super molecular structure of PHB/PANI nanofibres.

2.7. Poly (3, 4-ethylenedioxythiophene)

Although PPy and PANI have been the most extensively conductive polymers for tissue engineering and regenerative medicine, recently the potential of polythiophene conductive polymer for tissue engineering have been approved. This polymer has received significant attention due to a wide range of promising electronic and electrochemical applications [55,56]. PEDOT can be considered as the most successful polythiophene due to its specific characteristics [57-65]. PEDOT can also be considered as the most stable conducting polymer currently available because of not only high conductivity but also unusual environmental and electrochemical stabilities in the oxidized state [57-60]. Recently, Bolin *et al.* [66] reported electronically conductive and electrochemically active 3D-nanocomposite scaffolds based on electrospun poly(ethylene terephthalate) (PET) nanocomposite fibers. They employed vapour phase polymerization to achieve a uniform and conformal coating of PEDOT doped with tosylate on the nano-fibers. They observed that the PEDOT coatings had a large impact on the wettability, turning the hydrophobic PET fibers super-hydrophilic. According to Fig. 4, the SH-SY5Y cells adhered well and showed healthy morphology. These electrically active nanocomposite scaffolds were used to induce Ca^{2+} signalling in SH-SY5Y neuroblastoma cells. Their reported nanocomposite fibers represented a class of 3D host environments that combined excellent adhesion and proliferation for neuronal cells with the possibility to regulate their signalling.

2.8. Piezoelectric polymeric nanocomposites

Recent studies on the application of conductive materials showed that piezoelectric polymeric materials can also be considered for tissue engineering applications. Piezoelectric polymeric materials can generate surface charges by even small mechanical deformations [67]. Poly(vinylidenefluoride) (PVDF) is a synthetic, semicrystalline polymer with piezoelectric properties that can be potentially used for biomedical application due to their unique molecular structure [107]. An electrical charged porous nanocomposite could be a promising approach for a number of tissue engineering applications. Reported data on piezoelectric polymeric nanocomposites showed that after electrical stimulation, cellular interaction and tissue growth might be improved [68].

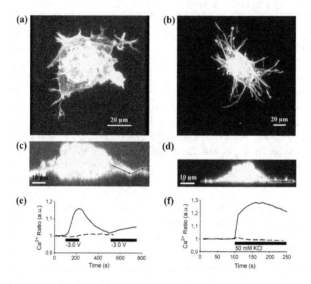

Figure 4. Confocal micrograph top view Y-axis projection of Tritc-phalloidin stained cluster of SH-SY5Y cells growing on (a) VPP-PEDOT coated nano-fiber (b) cell culture treated glass. Confocal micrograph side view Z-projection of Tritc-phalloidin stained cluster of SH-SY5Y cells growing on (c) VPP-PEDOT coated nano-fiber surface and (d) cell culture treated glass. Arrows indicate the direction of neurites. (e) Solid line shows intracellular Ca^{2+} flux in FURA-2-AM loaded SH-SY5Y cells cultured on nano-fiber surface. A potential of -3.0V is applied at 100 s. The potential is turned off at 250 s and turned on again at 500 s. Dashed line shows cell treated with 50µM nifedipine in order to block the VOCCs and stimulated with -3.0V at 100 s until 380 s. (f) Solid line shows intracellular Ca^{2+} flux in FURA-2-AM loaded SH-SY5Y cells cultured in cell culture dish 50mM KCl was added at 100 s. Dashed line shows cell treated with 50µM nifedipine in order to block the VOCCs and stimulated in the same way [66].

3. Applications of conducting polymers

3.1. Applications of conducting polymers: general view

Conductive polymers exhibit attractive properties such as ease of synthesis and processing [69]. The unique properties of this type of materials have recently given a wide range of applications in the biological field. Research on conductive polymers for biomedical applications expanded extrimly in the 1980s, and they were shown via electrical stimulation, to modulate cellular activities (e.g. cell adhesion, migration, DNA synthesis and protein secretion) [70-73]. Since then many studies have been done on nerve, bone, muscle, and cardiac cells. The unique characteristics of conducting polymers have been shown to be useful in many biomedical applications, specially tissue engineering nanocomposite scaffolds and drug delivery devices [74]. In comparison to other conductive materials for biological applications, conducting polymers are inexpensive, easy to synthesize, and versatile. In addition, conducting polymers permit control over the level and duration of electrical stimulation for tissue engineering applications.

3.2. Use and modification of conducting polymers for drug delivery

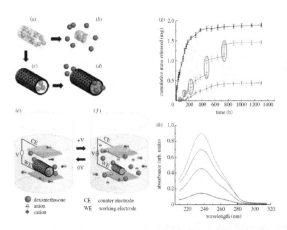

Figure 5. a) dexamethasone-loaded electrospun PLGA, (b) hydrolytic degradation of PLGA fibres leading to release of the drug and (c) and (d) electrochemical deposition of PEDOT around the dexamethasone-loaded PLGA fibre slows down the release of dexamethasone. (e) PEDOT nanotubes in a neutral electrical condition. (f) External electrical stimulation controls the release of dexamethasone from the PEDOT nanotubes. By applying a positive voltage, electrons are injected into the chains and positive charges in the polymer chains are compensated. (g) Cumulative mass release of dexamethasone from: PLGA nanoscale fibres (black squares), PEDOT-coated PLGA nanoscale fibres (red circles) without electrical stimulation and PEDOT-coated PLGA nanoscale fibres with electrical stimulation of 1 V applied at the five specific times indicated by the circled data points (blue triangles). (h) UV absorption of dexamethasone-loaded PEDOT nanotubes after 16 h (black), 87 h (red), 160 h (blue) and 730 h (green). [80]

Developing novel drug-delivery systems will open up new applications that were previously unsuited to traditional delivery systems. The use of conducting polymers in drug delivery is an excellent approach due to their biocompatibility and their possibility of using them in *in vivo* applications for real time monitoring of drugs in biological environments [75]. Controlled drug release can also be facilitated using a change in conductive polymer redox state to increase permeation of drugs such as dexamethasone [76]. Electrical stimulation of conductive polymers has been used to release a number of therapeutic proteins and drugs like nerve growth factor [77], dexamethasone [78] and heparin [79]. Another study demonstrated the use of PEDOT nanotubes polymerized on top of electrospun poly(lactic-co-glycolic acid) (PLGA) nanocomposite fibres for the potential release of the drug dexamethasone. Here, dexamethasone was incorporated within the PLGA nanocomposite fibres and then PEDOT was polymerized around the dexamethasone-loaded PLGA nanocomposite. As the PLGA fibres degraded, dexamethasone molecules remained inside the PEDOT nanotubes. These PEDOT nanotubes favoured controlled drug release upon electrical stimulation. Fig. 5 demonstrates the incorporation and release mechanism of dexamethasone from PEDOT nanotubes due to electrical stimulation. This drug-delivery system had the potential of immense interest for the treatment of cancer and tissue engineering and regenerative medicin [80].

3.3. Use and modification of conducting polymers for bioactuators

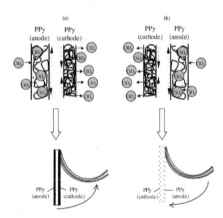

Figure 6. The triple layer device (polypyrrole(ClO$_4^-$)/non-conducting and adherent polymer/polypyrrole(ClO$_4^-$)) and its macroscopic movement produced as a consequence of volume change in the polypyrrole films. (a) A current flows and the left polypyrrole film acting as the anode is swelled by the entry of the hydrated counter ions (ClO$_4^-$). Simultaneously, the right film acting as the cathode contracts and shrinks because of the expulsion of the counter ions. These volume changes and the constant length of the non-conducting film promote the movement of the triple layer towards the polypyrrole film that is being contracted. (b) By changing the direction of the current, the movement takes place in the opposite direction. The muscle works in LiClO$_4$ aqueous solution [83]

Bioactuators are devices that are used to create mechanical force, which in turn can be used as artificial muscles. The phenomenon of change in the volume of the conducting polymers scaffold upon electrical stimulation has been employed in the construction of bioactuators. In artificial muscle applications, two layers of conducting polymers are placed in a triple layer arrangement, where the middle layer comprises a non-conductive material [81]. When current is applied across the two conducting polymers films, one of the films is oxidized and the other reduced. The oxidized film expands owing to the inflow of dopant ions, whereas the reduced film expels the dopant ions and in the process shrinks, as depicted in Fig. 6 [81]. Conducting actuators have many features that make them ideal candidates for artificial muscles, including that they:

- can be electrically controlled,
- have a large strain which is favourable for linear, volumetric or bending actuators,
- possess high strength,
- require low voltage for actuation (1 V or less),

- can be positioned continuously between minimum and maximum values,

- work at room/body temperature,

- can be readily microfabricated and are light weight, and

- can operate in body fluids [82].

3.4. Use and modification of conducting polymers for tissue engineering applications

The essential properties of conductive polymers desired for tissue engineering and regenerative medicine are conductivity, reversible oxidation, redox stability, biocompatibility, hydrophobicity, three-dimensional geometry and surface topography. Conductive polymers are widely used in tissue engineering due to their ability to subject cells to an electrical stimulation. Studies have addressed cell compatibility when a current or voltage is applied to PPy. An advantage offered by conducting polymers is that the electrochemical synthesis allows direct deposition of a polymer on the surface while simultaneously trapping the protein molecules [84].

In a recent study the release of NGF from PPy nanocomposites by using biotin as a co-dopant during the electrical polymerization was investigated [85]. In this research, NGF was biotinylated and immobilized to streptavidin entrapped within PPy nanocomposites doped with both biotin and dodecylbenzenesulfonate. The release of heparin from hydrogels immobilized onto PPy nanocomposites could also be triggered by electrical stimulation [86]. PVA hydrogels were covalently immobilized onto PPy via grafting of aldehyde groups to PPy and chemical reaction of these with hydroxyl groups from the hydrogel as shown in Fig. 7.

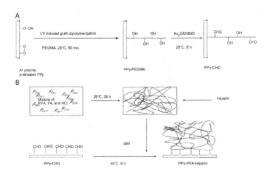

Figure 7. Controlled release of heparin from poly(vinyl alcohol) (PVA) hydrogels immobilized on PPy. (A) Post-polymerization of PPy to incorporate aldehyde groups. (B) Covalent immobilization of PVA hydrogels containing heparin on PPy substrates. Controlled release of heparin was obtained by electrical stimulation of PPy [148].

Electrically conducting polymers have attracted much interest for the construction of nerve guidance channels. The use of conducting polymers can help locally deliver electrical stimulus. It can also provide a physical template for cell growth and tissue repair and allow precise external control over the level and duration of stimulation [87,88]. The importance of conducting polymeric nanocomposites is based on the hypothesis that such composites can be used to host the growth of cells, so that electrical stimulation can be applied directly to the cells through the composite, proved to be beneficial in many regenerative medicine strategies, including neural and cardiac tissue engineering [89].

Recently, Li *et al.* [90] blended PANI with a natural protein, gelatin, and prepared nanocomposite fibrous scaffolds to investigate the potential application of such a blend as conductive scaffold for tissue engineering applications. As can be seen in Fig. 8, SEM analysis of the scaffolds containing less than 3% PANI in total weight, revealed uniform fibers with no evidence for phase egregation, as also confirmed by DSC.

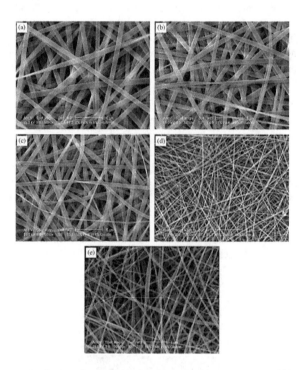

Figure 8. SEM micrographs of gelatin fibers (a) and PANi-gelatin blend fibers with ratios of (b) 15:85; (c) 30:70; (d) 45:55; and (e) 60:40. Original magnifications are 5000× for (a–d) and 20000× for (e). Figure shows the electrospun fibers were homogeneous while 60:40 fibers were electrospun with beads [90]

To test the usefulness of PANI/gelatin blends as a fibrous matrix for supporting cell growth, H9c2 rat cardiac myoblast cells were cultured on fiber-coated glass cover slips. Cell cultures were evaluated in terms of cell proliferation and morphology. According to Fig. 9, the results indicated that all PANI/gelatin blend fibers supported H9c2 cell attachment and proliferation to a similar degree as the control tissue culture-treated plastic (TCP) and smooth glass substrates.

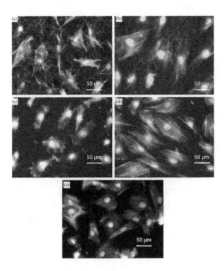

Figure 9. Morphology of H9c2 myoblast cells at 20 h of post-seeding on: (a) gelatin fiber; (b) 15:85 PANI/gelatin blend fiber; (c) 30:70 PANI/gelatin blend fiber; (d) 45:55 PANI/gelatin blend fibers; and (e) glass matrices. Staining for nuclei-bisbenzimide and actin cytoskeleton-phalloidin, fibersautofluorescence, original magnification 400× [90].

Depending on the concentrations of PANI, the cells initially displayed different morphologies on the fibrous substrates, but after 1week all cultures reached confluence of similar densities and morphology. Taken together they suggested that PANI/gelatin blend nanocomposite fibers could provide a novel conductive material well suited as biocompatible scaffolds for tissue engineering.

4. Conclusion

Tissue engineering is a new concept which is a growing area of research, in which cells are seeded on nanocomposite scaffolds and then implanted in defected part of body. Appropriate stimuli (chemical, biological, mechanical and electrical) can be applied and over a relatively short time new tissue can be formed to help restore function in the patient. The ideal scaffolds should have an appropriate surface chemistry and microstructures to facilitate cellular attachment, proliferation and differentiation. In addition, the scaffolds should possess adequate mechanical strength and biodegradation rate without any undesirable by-products. Among different materials, conducting polymers are one of the materials that can be employed to facilitate communication with neural system for regenerative purposes. In this chapter the recent methods of the synthesis of nanocomposite scaffolds using different conducting polymers was reviewed. The ability of conductive scaffolds to accept and modulate the growth of a few different cell types including endothelial, nerve, and chromaffin cells have shown a bright future in the field of tissue engineering and regenerative medicine.

Acknowledgements

This review chapter book is partially based upon work supported by Air Force Office of Scientific Research (AFOSR) High Temperature Materials program under grant no. FA9550-10-1-0010 and the National Science Foundation (NSF) under grant no. 0933763.

Author details

Masoud Mozafari[1], Mehrnoush Mehraien[1], Daryoosh Vashaee[2] and Lobat Tayebi[1*]

*Address all correspondence to: lobat.tayebi@okstate.edu

1 Helmerich Advanced Technology Research Center, School of Material Science and Engineering, Oklahoma State University, USA

2 Helmerich Advanced Technology Research Center, School of Electrical and Computer Engineering, Oklahoma State University, USA

References

[1] Mozafari, M., Moztarzadeh, F., Rabiee, M., Azami, M., Tahriri, M., Moztarzadeh, Z., & Nezafati, N. (2010). Development of Macroporous Nanocomposite Scaffolds of Gelatin/Bioactive Glass Prepared Through Layer Solvent Casting Combined with Lamination Technique for Bone Tissue Engineering. *Ceramics International*, 36, 2431-2439.

[2] Mozafari, M., Rabiee, M., Azami, M., & Maleknia, S. (2010). Biomimetic Formation of Apatite on the Surface of Porous Gelatin/Bioactive Glass Nanocomposite Scaffolds. *Applied Surface Science*, 257, 1740-1749.

[3] Mozafari, M., Moztarzadeh, F., Rabiee, M., Azami, M., Tahriri, M., & Moztarzadeh, Z. (2010). Development of 3D Bioactive Nanocomposite Scaffolds Made from Gelatin and Nano Bioactive Glass for Biomedical Applications. *Advanced Composites Letters*, 19, 91-96.

[4] Hamlekhan, A., Mozafari, M., Nezafati, N., Azami, M., & Hadipour, H. (2010). A Proposed Fabrication Method of Novel PCL-GEL-HAp Nanocomposite Scaffolds for Bone Tissue Engineering Applications. *Advanced Composites Letters*, 19, 123-130.

[5] Poursamar, S. A., Azami, M., & Mozafari, M. (2011). Controllable Synthesis and Characterization of Porous Polyvinyl Alcohol/Hydroxyapatite Nanocomposite Scaffolds via an in Situ Colloidal Technique. *Colloids and Surfaces B: Biointerfaces*, 84, 310-316.

[6] Hamlekhan, A., Moztarzadeh, F., Mozafari, M., Azami, M., & Nezafati, N. (2011). Preparation of Laminated Poly(ε-caprolactone)-Gelatin-Hydroxyapatite Nanocomposite Scaffold Bioengineered via Compound Techniques for Bone Substitution. *Biomatter*, 1, 1-11.

[7] Ghafari Nazari, A., & Mozafari, M. (2012). Simulation of Structural Features on Mechanochemical Synthesis of $Al_2O_3TiB_2$ Nanocomposite by Optimized Artificial Neural Network. *Advanced Powder Technology*, 23, 220-227.

[8] Hamlehkhan, A., Mozafari, M., Nezafati, N., Azami, M., & Samadikuchaksaraei, A. (2012). Novel Bioactive Poly(ε-caprolactone)-Gelatin-Hydroxyapatite Nanocomposite Scaffolds for Bone Regeneration. *Key Engineering Materials*, 493-494, 909-915.

[9] Baghbani, F., Moztarzadeh, F., Gafari Nazari, A., Razavi Kamran, A. H., Tondnevis, F., Nezafati, N., Gholipourmalekabadi, M., & Mozafari, M. (2012). Biological Response of Biphasic Hydroxyapatite/Tricalcium Phosphate Scaffolds Intended for Low Load-Bearing Orthopaedic Applications. *Advanced Composites Letters*, 21, 16-24.

[10] Jalali, N., Moztarzadeh, F., Mozafari, M., Asgari, S., Motevalian, M., & Naghavi Alhosseini, S. (2012). Surface Modification of Poly(lactide-co-glycolide) Nanoparticles by d-α-tocopheryl Polyethylene Glycol 1000 Succinate as Potential Carrier for the Delivery of Drugs to the Brain. *Colloids and Surfaces A: Physicochemical and Engineering Aspects*, 392, 335-342.

[11] Pomfret, S. J., Adams, P. N., Comfort, N. P., et al. (2000). Electrical and Mechanical Properties of Polyaniline Fibres Produced by a One-Step Wet Spinning. *Polymer*, 41, 2265-2269.

[12] Yu, Q. Z., Shi, M. M., Deng, M., et al. (2008). Morphology and Conductivity of Polyaniline Sub-Micron Fibers Prepared by Electrospinning. *Mater Sci Eng B*, 150, 70-76.

[13] Veluru, J. B., Satheesh, K. K., Trivedi, D. C., et al. (2007). Electrical Properties of Electrospun Fibers of PANI-PMMA Composites. *J Eng Fibers Fabrics*, 2, 25-31.

[14] Bishop, A., & Gouma, P. (2005). Leucoemeraldine Based Polyaniline-Poly-Vinylpyrrolidone Electrospun Composites and Bio-Composites: A Preliminary Study of Sensing Behavior. *Rev Adv Mater Sci*, 10, 209-214.

[15] Desai, K., Lee, J. S., & Sung, C. (2004). Nanocharacterization of Electrospun Nanofibers of Polyaniline/Poly Methyl Methacrylate Blends Using SEM, TEM and AFM. *Microsc Microanal*, 10, 556-557.

[16] Ju, Y. W., Park, J. H., Jung, H. R., et al. (2007). Electrochemical Properties of Polypyrrole/Sulfonated SEBS Composite Nanofibers Prepared by Electrospinning. *Electrochim Acta*, 52, 4841-4847.

[17] Schmidt, E., Shastri, V. R., Vacanti, J. P., & Langer, R. (1997). *Proc. Natl. Acad. Sci. Usa*, 94, 8948.

[18] De Giglio, E., Sabbatini, L., & Zambonin, P. G. J. (1999). *Biomater. Sci. Polym. Ed*, 10, 845.

[19] Garner, B., Georgevich, A., Hodgson, A. J., Liu, L., & Wallace, G. G. J. (1999). *Biomed. Mater. Res*, 44, 121.

[20] Valentini, R. F., Vargo, T. G., Gardellajr, J. A., & Aebischer, P. (1992). *Biomaterials*, 13, 193.

[21] Kotwal, A., & Schmidt, C. E. (2001). *Biomaterials*, 22, 1055.

[22] Xu, L. B., Chen, W., Mulchandani, A., & Yan, Y. (2005). Reversible Conversion of Conducting Polymer Films from Superhydrophobic to Superhydrophilic. . Angew. Chem. Int. Ed doi:10.1002/anie.200500868 , 44, 6009-6012.

[23] Rajeswari, R., Subramanian, S., Jayarama, R. V., Shayanti, M., & Seeram, R. (2010). Applications of Conducting Polymers and Their Issues in Biomedical Engineering. J. R. Soc. Interface published online 7 July, doi: 10.1098/rsif.2010.0120.focus.

[24] Street, G. B. (1986). Polypyrrole: from Powders to Plastics. In: Skotheim T. A., editor. *Handbook of conducting polymers*, vol. I. New York: Marcel Dekker, 265-291.

[25] Shirakawa, H., Louis, E. J., MacDiarmid, A. G., Chiang, C. K., & Heeger, A. J. (1977). Synthesis of Electrically Conducting Organic Polymers: Halogen Derivatives of Poly-acetylene, (CH)x. *J Chem Soc Chem Commun*, 578-580.

[26] Heeger, A. J. (2001). Semiconducting and Metallic Polymers: the Fourth Generation of Polymeric Materials (Nobel Lecture). *Angew Chem Int Ed*, 40, 2591-2611.

[27] Feast, W. J. (1986). Synthesis of Conducting Polymers. In: Skotheim T. A., editor. *Handbook of conducting polymers*, vol. I. New York: Marcel Dekker, 1-43.

[28] Hong, S. Y., & Marnick, D. S. (1992). Understanding the Conformational Stability and Electronic Structures of Modified Polymers Based on Polythiophene. *Macromolecules*, 4652-4657.

[29] Guimard, N. K., Gomez, N., & Schmidt, C. E. (2007). Conducting Polymers in Bio-medical Engineering. *Prog. Polym. Sci.*, 32, 876-921.

[30] Green, R. A., Baek, S., Poole-Warren, L. A., & Martens, P. J. (2010). Conducting Poly-mer-Hydrogels for Medical Electrode Applications. *Sci. Technol. Adv. Mater*, 11 014107 (13pp), doi: 10.1088/1468-6996/11/1/014107.

[31] Kamalesh, S., Tan, P., Wang, J., Lee, T., Kang, E. T., & Wang, C. H. (2000). *J. Biomed. Mater. Res.*, 52, 467.

[32] Wang, H. J., Ji, L. W., Li, D. F., & Wang, J. Y. (2008). *J. Phys. Chem. B*, 112, 2671.

[33] Shreyas, S. R., & Jessica, W. (2009). *Front. Neuroeng.*, 2, 6.

[34] Cogan, S. F. (2008). *Annu. Rev. Biomed. Eng.*, 10, 275.

[35] Li, G. C., & Pickup, P. G. (2000). *PCCP Phys. Chem. Chem. Phys.*, 2, 1255.

[36] Tourillon, G., & Garnier, F. (1983). *J. Electrochem. Soc.*, 130, 2042.

[37] Cui, X., & Martin, D. C. (2003). *Sensors Actuators B*, 89, 92.

[38] Yamato, H., Ohwa, M., & Wernet, W. J. (1995). *Electroanal. Chem.*, 397, 163.

[39] Ghasemi-Mobarakeh, L., Prabhakaran, M. P., Morshed, M., Nasr Esfahani, M. H., Ba-harvand, H., Kiani, S., & Al-Deyab, S. S. (2011). Ramakrishna S. Application of Con-ductive Polymers, Scaffolds and Electrical Stimulation for Nerve Tissue Engineering. *J Tissue Eng Regen Med*, 5, 17-35.

[40] Naghavi Alhosseini, S., Moztarzadeh, F., Mozafari, M., Asgari, S., Dodel, M., Samadi-kuchaksaraei, A., Kargozar, S., & Jalali, N. (2011). Synthesis and Characterization of Electrospun Polyvinyl Alcohol Nanofibrous Scaffolds Modified by Blending with Chitosan for Neural Tissue Engineering. *International Journal of Nanomedicine*, 6, 1-10.

[41] Otero, T. F., & Sansinena, J. M. (1998). Soft and Wet Conducting Polymers for Artifi-cial Muscles. *Adv. Mater.*, 10, 491-494.

[42] Abidian, M. R., Kim, D. H., & Martin, D. C. (1998). Conducting Polymer Nanotubes for Controlled Drug Release. *Adv. Mater.*, 18, 405-409, doi:10.1002/adma.200501726.

[43] Abidian, M. R., Ludwig, K. A., Marzullo, T. C., Martin, D. C., & Kipke, D. R. (2009). Interfacing Conducting Polymer Nanotubes with the Central Nervous System: Chronic Neural Recording Using poly(3,4-ethylenedioxythiophene) Nanotubes. *Adv. Mater.*, 21, 3764-3770, doi:10.1002/adma.200900887.

[44] Schmidt, C.E., Shastri, V. R., Vacanti, J. P., & Langer, R. (1997). Stimulation of Neurite Outgrowth Using an Electrically Conducting Polymer. *Proc. Natl Acad. Sci. USA*, 94, 8948-8953.

[45] Warren, L. F., Walker, J. A., Anderson, D. P., Rhodes, C. G., & Buckley, L. J. (1989). A study of Conducting Polymer Morphology. *J. Electrochem. Soc.*, 136, 2286-2295.

[46] Breads, J. L., & Silbey, R. (1991). Conjugated Polymers. Kluwer Academic: Amster-dam, The Netherlands.

[47] Wong, J. Y., Langer, R., & Ingberi, D. E. (1994). Electrically Conducting Polymers Can Noninvasively Control the Shape and Growth of Mammalian Cells. *Proc Natl Acad Sci USA*, 91, 3201-3204.

[48] Sanchvi, A. B., Miller, K. P. H., Belcher, A. M., et al. (2005). Biomaterials Functionali-zation Using a Novel Peptide That Selectively Binds to a Conducting Polymer. *Nat Mater*, 4, 496-502.

[49] Huang, L., Hu, J., Lang, L., et al. (2007). Synthesis and Characterization of Electroac-tive and Biodegradable ABA Block Copolymer of Polylactide and Aniline Pentamer. *Biomaterials*, 28, 1741-1751.

[50] Wong, J. Y., Langer, R., & Ingber, D. E. (1994). Electrically Conducting Polymers Can Noninvasively Control the Shape and Growth of Mammalian Cells. *Proc Natl Acad Sci USA*, 91, 3201-3204.

[51] Moroder, P., Runge, M. B., Wang, H., Ruesink, T., Lu, L., Spinner, R. J., Windebank, A. J., & Yaszemski, M. J. (2011). Material Properties and Electrical Stimulation Regimens of Polycaprolactone Fumarate-Polypyrrole Scaffolds as Potential Conductive Nerve Conduits. *Acta Biomaterialia*, 7, 944-953.

[52] Nalwa, H. S. (1997). Handbook of Organic Conductive Molecules and Polymers. Wiley: New York.

[53] Mattioli-Belmonte, M., Giavaresi, G., Biagini, G., et al. (2003). Tailoring Biomaterial Compatibility: in Vivo Tissue Response Versus in Vitro Cell Behavior. *Int J Artif Organs*, 26, 1077-1085.

[54] Fryczkowski, R., & Kowalczyk, T. (2009). Nanofibres from Polyaniline/Polyhydroxybutyrate Blends. *Synthetic Metals*, 159, 2266-2268.

[55] Crispin, X., Marciniak, S., Osikowicz, W., Zotti, G., van Der Gon, A. W. D., Louwt, F., et al. (2003). *J Polym Sci Pol Phys*, 41, 2561.

[56] Sarac, A. S., Sonmez, G., & Cebeci, F. C. (2003). *J Appl Electrochem*, 33, 295.

[57] Breiby, D. W., Samuelsen, E. J., Groenendaal, L. B., & Struth, B. (2003). *J Polym Sci Pol Phys*, 41, 945.

[58] Jonsson, S. K. M., Birgersson, J., Crispin, X., Grezynsky, G., Osikowicz, W., van der Gon, A. W. D., et al. (2003). *Synthetic Met*, 139, 1.

[59] Ocampo, C., Oliver, R., Armelin, E., Alema'n, C., & Estrany, F. (2006). *J Polym Res*, 13, 193.

[60] Liesa, F., Ocampo, C., Alema'n, C., Armelin, E., Oliver, R., & Estrany, F. (2006). *J Appl Polym Sci*, 102, 1592.

[61] Marsella, M. J., & Reid, R. (1999). *J. Macromolecules*, 32, 5982.

[62] Otero, T. F., & Cortes, M. T. (2003). *Adv Mater*, 15, 279.

[63] Yu, H. H., Xu, B., & Swager, T. M. (2003). *J Am Chem Soc*, 125, 1142.

[64] Casanovas, J., Zanuy, D., & Alema'n, C. (2006). *Angew Chem Int Edit*, 45, 1103.

[65] Kros, A., Van Hovell, S. W. F. M., Sommerdijk, N. A. J. M., & Nolte, R. J. M. (2001). *Adv Mater*, 13, 1555.

[66] Bolin, M. H., Svennersten, K., Wang, X., Chronakis, I. S., Richter-Dahlfors, A., Jager, E. W. H., & Berggren, M. (2009). Nano-Fiber Scaffold Electrodes Based on PEDOT for Cell Stimulation. *Sensors and Actuators B*, 142, 451-456.

[67] Valentini, R. F., Vargo, T. G., Gardella, J. A., et al. (1992). Electrically Charged Polymeric Substrates Enhance Nerve Fibre Outgrowth in Vivo. *Biomaterials*, 13, 183-190.

[68] Weber, N., Lee, Y. S., Shanmugasundaram, S., Jaffe, M., & Arinzeh, T. L. (2010). Characterization and in Vitro Cytocompatibility of Piezoelectric Electrospun Scaffolds. *Acta Biomaterialia*, 6, 3550-3556.

[69] Heeger, A. J. (2002). Semiconducting and Metallic Polymers: the Fourth Generation of Polymeric Materials. *Synth Met*, 125, 23-42.

[70] Foulds, N. C., & Lowe, C. R. (1986). Enzyme Entrapment in Electrically Conducting Polymers. *J Chem Soc Faraday Trans*, 82, 1259-1264.

[71] Umana, M., & Waller, J. (1986). Protein Modified Electrodes: the Glucose/Oxidase/Polypyrrole System. *Anal Chem*, 58, 2979-2983.

[72] Venugopal, J., Molamma, P., Choon, A. T., Deepika, G., Giri Dev, V. R., & Ramakrishna, S. (2009). Continuous Nanostructures for the Controlled Release of Drugs. *Curr. Pharm. Des.*, 15, 1799-1808.

[73] Adeloju, S. B., & Wallace, G. G. (1996). Conducting Polymers and the Bioanalytical Sciences: New Tools for Biomolecular Communication. *A review. Analyst*, 121, 699-703.

[74] Harwood, G. W. J., & Pouton, C. W. (1996). Amperometric Enzyme Biosensors for the Analysis of Drug and Metabolites. *Adv. Drug. Deliv. Rev.*, 18, 163-191, doi: 10.1016/0169-409X(95)00093-M.

[75] Stassen, I., Sloboda, T., & Hambitzer, G. (1995). Membrane with Controllable Permeability for Drugs. *Synth. Met.*, 71, 2243-2244, doi:10.1016/0379-6779(94)03241-W.

[76] Pernaut, J. M., & Reynolds, J. R. (2000). Use of Conducting Electroactive Polymers for Drug Delivery and Sensing of Bioactive Molecules. A Redox Chemistry Approach. *J. Phys. Chem. B*, 104, 4080-4090, doi:10.1021/jp994274o).

[77] Hodgson, A. J., John, M. J., Campbell, T., Georgevich, A., Woodhouse, S., & Aoki, T. (1996). Integration of Biocomponents with Synthetic Structures: Use of Conducting Polymer Polyelectrolyte Composites. *Proc. SPIE. Int. Soc. Opt. Eng.*, 2716, 164-176, doi: 10.1117/12.232137.

[78] Wadhwa, R., Lagenaur, C. F., & Cui, X. T. (2006). Electrochemically Controlled Release of Dexamethasone from Conducting Polymer Polypyrrole Coated Electrode. *J. Control. Rel.*, 110, 531-541, doi: 10.1016/j.jconrel.2005.10.027.

[79] Li, Y., Neoh, K. G., Cen, L., & Kang, E. T. (2005a). Controlled Release of Heparin from Polypyrrole-Poly(Vinyl Alcohol) Assembly by Electrical Stimulation. *J. Biomed. Mater. Res. A*, 73A, 171-181, doi:10.1002/jbm.a.30286.

[80] Abidian, M. R., Kim, D. H., & Martin, D. C. (2006). Conducting-Polymer Nanotubes for Controlled Drug Release. *Adv. Mater.*, 18, 405-409.

[81] Otero, T. F., & Cortes, M. T. (2003). A Sensing Muscle. *Sens. Actuat. B*, 96, 152-156, doi: 10.1016/S0925-4005(03)00518-5.

[82] Smela, E. (2003). Conjugated Polymer Actuators for Biomedical Applications. *Adv. Mater.*, 15, 481-494, doi: 10.1002/ adma.200390113.

[83] Otero, T. F., & Cortés, M. T. (2003). A Sensing Muscle. *Sensors and Actuators B*, 96, 152-156.

[84] Bartlett, P. N., & Whitaker, R. G. (1988). Modified Electrode Surface in Amperometric Biosensors. *Med. Biol. Eng. Comput.*, 28, 10-17, doi:10.1007/BF02442675.

[85] George, P. M., LaVan, D. A., Burdick, J. A., Chen, C. Y., Liang, E., & Langer, R. (2006). Electrically Controlled Drug Delivery from Biotindoped Conductive Polypyrrole. *Adv Mater*, 18, 577-581.

[86] Li, Y., Neoh, K. G., & Kang, E. T. (2005). Controlled Release of Heparin from Polypyr-role-Poly(Vinyl Alcohol) Assembly by Electrical Stimulation. *J Biomed Mater Res A*, 73A, 171-181.

[87] Chronakis, I. S., Grapenson, S., & Jakob, A. (2006). Conductive Polypyrrole Nanofib-ers via Electrospinning: Electrical and Morphological Properties. *Polymer*, 47, 1597-1603.

[88] Zhang, Q., Yan, Y., Li, S., et al. (2010). The Synthesis and Characterization of a Novel Biodegradable and Electroactive Polyphosphazene for Nerve Regeneration. *Mater Sci Eng C*, 30, 160-166.

[89] Bettinger, C. J., Bruggeman, J. P., Misra, A., et al. (2009). Biocompatibility of Biode-gradable Semiconducting Melanin Films for Nerve Tissue Engineering. *Biomaterials*, 30, 3050-3057.

[90] Li, M., Guo, Y., Wei, Y., MacDiarmid, A. G., & Lelkes, P. I. (2006). Electrospinning Polyaniline-Contained Gelatin Nanofibers for Tissue Engineering Applications. *Bio-materials*, 27, 2705-2715.

[91] Buijtenhuijs, P., Buttafoco, L., Poot, A. A., Daamen, W. F., van Kuppevelt, T. H., Dijk-stra, P. J., et al. (2004). Tissue Engineering of Blood Vessels: Characterization of Smooth-Muscle Cells for Culturing on Collagen-and-Elastinbased Scaffolds. *Biotech-nol Appl Biochem*, 39, (Pt 2) 141-149.

[92] Lu, Q., Ganesan, K., Simionescu, D. T., & Vyavahare, N. R. (2004). Novel Porous Aortic Elastin and Collagen Scaffolds for Tissue Engineering. *Biomaterials*, 25(22), 5227-5237.

[93] Tan, K. H., Chua, C. K., Leong, K. F., Naing, M. W., & Cheah, C. M. (2005). Fabrica-tion and Characterization of Three-Dimensional Poly(ether-ether-ketone)-Hydroxya-patite Biocomposite Scaffolds Using Laser Sintering. *Proc Inst Mech Eng [H]*, 219(3), 183-194.

[94] Di Martino, A., Sittinger, M., & Risbud, M. V. (2005). Chitosan: a Versatile Biopoly-mer for Orthopaedic Tissue-Engineering. *Biomaterials*, [Epub ahead of print].

[95] Kim, K., Yu, M., Zong, X., Chiu, J., Fang, D., Seo, Y. S., et al. (2003). Control of Degradation Rate and Hydrophilicity in Electrospun Non-Woven Poly(D,L-lactide) Nanofiber Scaffolds for Biomedical Applications. *Biomaterials*, 24, 4977-4985.

[96] Zong, X., Bien, H., Chung, C. Y., Yin, L., Fang, D., Hsiao, B. S., et al. (2005). Electrospun Fine-Textured Scaffolds for Heart Tissue Constructs. *Biomaterials*, 26(26), 5330-5338.

[97] Metzke, M., O'Connor, N., Maiti, S., Nelson, E., & Guan, Z. (2005). Saccharidepeptide Hybrid Copolymers as Biomaterials. *Angew Chem Int Ed Engl*, 44(40), 6529-6533.

[98] Boland, E. D., Matthews, J. A., Pawlowski, K. J., Simpson, D. G., Wnek, G. E., & Bowlin, G. L. (2004). Electrospinning Collagen and Elastin: Preliminary Vascular Tissue Engineering. *Front Biosci*, 9, 1422-1432.

[99] Li, M., Mondrinos, M. J., Gandhi, M. R., Ko, F. K., Weiss, A. S., & Lelkes, P. I. (2005). Electrospun Protein Fibers as Matrices for Tissue Engineering. *Biomaterials*, 26(30), 5999-6008.

[100] Rho, K. S., Jeong, L., Lee, G., Seo, B. M., Park, Y. J., Hong, S. D., et al. (2005). Electrospinning of Collagen Nanofibers: Effects on the Behavior of Normal Human Keratinocytes and Early Stage Wound Healing. *Biomaterials*, [Epub ahead of print].

[101] Riboldi, S. A., Sampaolesi, M., Neuenschwander, P., Cossu, G., & Mantero, S. (2005). Electrospun Degradable Polyesterurethane Membranes: Potential Scaffolds for Skeletal Muscle Tissue Engineering. *Biomaterials*, 26(22), 1606 1615.

[102] Ma, Z., Kotaki, M., Inai, R., & Ramakrishna, S. (2005). Potential of Nanofiber Matrix as Tissue-Engineering Scaffolds. *Tissue Eng*, 11(1-2), 101-109.

[103] Yang, F., Murugan, R., Wang, S., & Ramakrishna, S. (2005). Electrospinning of Nano/Micro Scale Poly(L-lactic acid) Aligned Fibers and Their Potential in Neural Tissue Engineering. *Biomaterials*, 26(15), 2603-2610.

[104] Khil, M. S., Bhattarai, S. R., Kim, H. Y., Kim, S. Z., & Lee, K. H. (2005). Novel Fabricated Matrix via Electrospinning for Tissue Engineering. *J Biomed Mater Res B Appl Biomater*, 72(1), 117-124.

[105] Khil, M. S., Cha, D. I., Kim, H. Y., Kim, I. S., & Bhattarai, N. (2003). Electrospun Nanofibrous Polyurethane Membrane as Wound Dressing. *J Biomed Mater Res B Appl Biomater*, 67(2), 675-679.

[106] Zeng, J., Yang, L., Liang, Q., Zhang, X., Guan, H., Xu, X., et al. (2005). Influence of the Drug Compatibility with Polymer Solution on the Release Kinetics of Electrospun Fiber Formulation. *J Control Release*, 105(1-2), 43-51.

[107] Buttafoco, L., Kolkman, N. G., Poot, A. A., Dijkstra, P. J., Vermes, I., & Feijen, J. (2005). Electrospinning Collagen and Elastin for Tissue Engineering Small Diameter Blood Vessels. *J Control Release*, 101(1-3), 322-324.

[108] MacDiarmid, A. G. (2001). Nobel lecture: Synthetic Metals: a Novel Role for Organic Polymers. *Rev Mod Phys*, 73, 701-712.

[109] Pedrotty, D. M., Koh, J., Davis, B. H., Taylor, D. A., Wolf, P., & Niklason, L. E. (2005). Engineering Skeletal Myoblasts: Roles of Three-Dimensional Culture and Electrical Stimulation. *Am J Physiol Heart Circ Physiol*, 288(4), H1620-1626.

[110] Azioune, A., Slimane, A. B., Hamou, L. A., Pleuvy, A., Chehimi, M. M., Perruchot, C., et al. (2004). Synthesis and Characterization of Active Esterfunctionalized Polypyrrole-Silica Nanoparticles: Application to the Covalent Attachment of Proteins. *Langmuir*, 20(8), 3350-3356.

[111] Arslan, A., Kiralp, S., Toppare, L., & Yagci, Y. (2005). Immobilization of Tyrosinase in Polysiloxane/Polypyrrole Copolymer Matrices. *Int J Biol Macromol*, 35(3-4), 163-167.

[112] Kim, D. H., Abidian, M., & Martin, D. C. (2004). Conducting Polymers Grown in Hydrogel Scaffolds Coated on Neural Prosthetic Devices. *J Biomed Mater Res A*, 71(4), 577-585.

[113] Kotwal, A., & Schmidt, C. E. (2001). Electrical Stimulation Alters Protein Adsorption and Nerve Cell Interactions with Electrically Conducting Biomaterials. *Biomaterials*, 22(10), 1055-1064.

[114] Sanghvi, A. B., Miller, K. P., Belcher, A. M., & Schmidt, C. E. (2005). Biomaterials Functionalization Using a Novel Peptide That Selectively Binds to a Conducting Polymer. *Nat Mater*, 4(6), 496-502.

[115] Lakard, S., Herlem, G., Valles-Villareal, N., Michel, G., Propper, A., Gharbi, T., et al. (2005). Culture of Neural Cells on Polymers Coated Surfaces for Biosensor Applications. *Biosens Bioelectron*, 20(10), 1946-1954.

[116] George, P. M., Lyckman, A. W., La Van , D. A., Hegde, A., Leung, Y., Avasare, R., et al. (2005). Fabrication and Biocompatibility of Polypyrrole Implants Suitable for Neural Prosthetics. *Biomaterials*, 26(17), 3511-3519.

[117] Wan, Y., Wu, H., & Wen, D. (2004). Porous-Conductive Chitosan Scaffolds for Tissue Engineering. 1. Preparation and Characterization. *Macromol Biosci*, 4(9), 882-890.

[118] Jiang, X., Marois, Y., Traore, A., Tessier, D., Dao, L. H., Guidoin, R., et al. (2002). Tissue Reaction to Polypyrrole-Coated Polyester Fabrics: an in Vivo Study in Rats. *Tissue Eng*, 8(4), 635-647.

[119] Bidez, P. R., Li, S., MacDiarmid, A. G., Venancio, E. C., Wei, Y., & Lelkes, P. I. (2006). Polyaniline, an Electroactive Polymer with Potential Applications in Tissue Engineering. *J Biomater Sci Polym*, 17(1-2), 199-212.

[120] Kamalesh, S., Tan, P., Wang, J., Lee, T., Kang, E. T., & Wang, C. H. (2000). Biocompatibility of Electroactive Polymers in Tissues. *J Biomed Mat Res*, 52(3), 467-478.

[121] Ahmad, N., & MacDiarmid, A. G. (1996). Inhibition of Corrosion of Steels with the Exploitation of Conducting Polymers. *Synth Met*, 78, 103-110.

[122] Yang, Y., Westerweele, E., Zhang, C., Smith, P., & Heeger, A. J. (1995). Enhanced Performance of Polymer Light-Emitting Diodes Using High-Surface Area Polyaniline Network Electrodes. *J Appl Phys*, 77, 694-698.

[123] MacDiarmid, A. G., Yang, L. S., Huang, W. S., & Humphrey, B. D. (1987). Polyaniline: Electrochemistry and Application to Rechargeable Batteries. *Synth Met*, 18, 393-398.

[124] Karyakin, A. A., Bobrova, O. A., Lukachova, L. V., & Karyakina, E. E. (1996). Potentiometric Biosensors Based on Polyaniline Semiconductor Films. *Sensors Actuators, B: Chem.*, B33(1-3), 34-38.

[125] Wei, Y., Lelkes, P. I., MacDiarmid, A. G., Guterman, E., Cheng, S., Palouian, K., et al. (2004). Electroactive Polymers and Nanostructured Materials for Neural Tissue Engineering. In: Qi-Feng Z., Cheng S. Z. D., editors. *Contemporary Topics in Advanced Polymer Science and Technology*, Beijing, China: Peking University Press, 430-436.

Chapter 4

Application of Nanocomposites for Supercapacitors: Characteristics and Properties

Dongfang Yang

Additional information is available at the end of the chapter

1. Introduction

Supercapacitors, ultracapacitors or electrochemical capacitors (ECs), are energy storage devices that store energy as charge on the electrode surface or sub-surface layer, rather than in the bulk material as in batteries, therefore, they can provide high power due to their ability to release energy more easily from surface or sub-surface layer than from the bulk. Since charging-discharging occurred on the surface, which does not induce drastic structural changes upon electroactive materials, supercapacitors possess excellent cycling ability. Due to those unique features, supercapacitors are regarded as one of the most promising energy storage devices. There are two types of supercapacitors: electrochemical double layer capacitors (EDLCs) and pseudocapacitors. In EDLCs, the energy is stored electrostatically at the electrode–electrolyte interface in the double layer, while in pseudocapacitors charge storage occurs via fast redox reactions on the electrode surface. There are three major types of electrode materials for supercapacitors: carbon-based materials, metal oxides/hydroxides and conducting polymers. Carbon-based materials such as activated carbon, mesoporous carbon, carbon nanotubes, graphene and carbon fibres are used as electrode active materials in EDLCs, while conducting polymers such as polyaniline, polypyrrole and polythiophene or metal oxides such as MnO_2, V_2O_5, and RuO_2 are used for pseudocapacitors. EDLCs depends only on the surface area of the carbon-based materials to storage charge, therefore, often exhibit very higher power output and better cycling ability. However, EDLCs have lower energy density values than pseudocapacitors since pseudocapacitors involve redox active materials to store charge both on the surface as well as in sub-surface layer.

Although carbon-based materials, metal oxides/hydroxides and conducting polymers are the most common electroactive materials for supercapacitor, each type of material has its own unique advantages and disadvantages, for example, carbon-based materials can provide high power density and long life cycle but its small specific capacitance (mainly double layer capac‐ itance) limits its application for high energy density devices. Metal oxides/hydroxides possess pseudocapacitance in additional to double layer capacitance and have wide charge/discharge potential range; however, they have relatively small surface area and poor cycle life. Conduct‐ ing polymers have the advantages of high capacitance, good conductivity, low cost and ease of fabrication but they have relatively low mechanical stability and cycle life. Coupling the unique advantages of these nano-scale dissimilar capacitive materials to form nanocomposite electroactive materials is an important approach to control, develop and optimize the struc‐ tures and properties of electrode material for enhancing their performance for supercapaci‐ tors. The properties of nanocomposite electrodes depend not only upon the individual components used but also on the morphology and the interfacial characteristics. Recently, con‐ siderable efforts have been placed to develop all kinds of nanocomposite capacitive materials, such as mixed metal oxides, conducting polymers mixed with metal oxides, carbon nanotubes mixed with conducting polymers, or metal oxides, and graphene mixed with metal oxides or conducting polymers. Design and fabrication of nanocomposite electroactive materials for su‐ percapacitors applications needs the consideration of many factors, such as material selection, synthesis methods, fabrication process parameters, interfacial characteristics, electrical con‐ ductivity, nanocrystallite size, and surface area, etc. Although significant progress has been made to develop nanocomposite electroactive materials for supercapacitor applications, there are still a lot of challenges to be overcome. This chapter will summarize the most recent devel‐ opment of this new area of research including the synthesis methods currently used for prepar‐ ing nanocomposite electroactive materials, types of nanocomposite electroactive materials investigated,structural and electrochemical characterization of nanocomposites, unique ca‐ pacitive properties of nanocomposite materials, and performance enhancement of nanocom‐ posite electroactive materials and its mechanism.

2. Fabrication and characterization of nanocomposite active electrode materials

2.1. Fabrication methods

To prepare mixed metal oxide nanocomposites, various synthesis methods including solid state reactions (i.e. thermal decomposition of mechanical mixtures of metal salts), mechani‐ cal mixing of metal oxides (i.e. ball milling), and chemical co-precipitation and electrochemi‐ cal anodic deposition from solutions containing metal salts, have been used. For example (examples in section 2.1 will be described in more details in section 3 and the references will also be given in section 3), Mn-Pb and Mn-Ni mixed oxide nanocomposites were prepared by reduction of $KMnO_4$ with Pb(II), and Ni(II) salts to form amorphous mixed oxide precipi‐ tant. Mn-V-W oxide, and Mn-V-Fe oxide were then directly deposited by anodic deposition on conductive substrates from aqueous solution consisting of mixed metal salts. Directly anodizing Ti–V alloys in ethylene glycol with HF electrolyte was used to synthesis mixed

V_2O_5–TiO_2 nanotube arrays. Hydrothermal process was also used to prepare SnO_2–Al_2O_3 mixed oxide nanocomposites involving urea as the hydrolytic agent in an autoclave.

Carbon nanotubes (CNTs)-metal oxide nanocomposites were prepared by either mechanically mixing CNTs with metal oxides in a mortar, or depositing metal oxides directly on CNTs by metal-organic chemical vapour deposition (CVD), wet-chemical precipitation, or electrochemically deposition. For example, IrO_2 nanotubes were deposited on multiwall CNTs using metal-organic CVD with the iridium source of $(C_6H_7)(C_8H_{12})Ir$ at 350∘C to form IrO_2-CNTS nanocomposite. The CNTs themselves were initially grown on stainless steel plate using thermal CVD. The MnO_2-CNTs nanocomposites were synthesized by direct current anodic deposition of MnO_2 from the $MnSO_4$ solution over electrophoretically deposited CNTs on the Ni substrate. RuO_2-CNTs was formed by impregnating CNTs with a ruthenium nitrosylnitrate solution and then followed by heat treatment to form composite electrode.

Nanocomposite of a conducting polymer with metal oxide, CNTs or graphene (GN) were mainly synthesized by in situ polymerization in solutions containing monomers of the conducting polymer and suspension of CNTs, metal oxide nanoparticles or GN nanosheets. For example, CNTs– polyaniline (PANI) nanocomposite was prepared from a solution consisting of CNTs and aniline monomer. With addition of an oxidant solution containing $(NH_4)_2S_2O_8$, polymerization of aniline on the surface of CNTs occurred to form CNTs–PANI nanocomposite. MoO_3-Poly 3,4-ethylenedioxythiophene (PEDOT) nanocomposites was synthesized by adding 3,4-ethylenedioxythiophene monomer into a lithium molybdenum nanoparticle suspension, and subsequently, Iron (III) chloride ($FeCl_3$) was added to the suspension as the oxidizing agent under microwave hydrothermal conditions for polymerization to occur. GN-PEDOT nanocomposite was chemically synthesized by oxidative polymerization of ethylene dioxythiophene using ammonium peroxydisulfate $[(NH_4)_2S_2O_8)]$ and $FeCl_3$ as oxidizing agents in a solution containing sodium polystyrene sulfonate Na salt, HCl, EDOT monomer and GN. G–PANI nanocomposite was chemically synthesized by oxidative polymerization of aniline monomers using ammonium peroxydisulfate $[(NH_4)_2S_2O_8)]$ in solution containing GN.

GN-metal oxide nanocomposites were prepared by chemical precipitation of metal oxide in the presence of GN nanosheets in the solution. For example, GN-CeO_2 nanocomposite was prepared by adding KOH solution dropwise into a $Ce(NO_3)_3$ aqueous solution in the presence of 3D GN material, followed by filtering, and drying. The GN-SnO_2/CNTs nanocomposite was synthesized by ultrasonicating the mixture of chemically functionalized GN and SnO_2–CNTs in water. Normally, GN sheets were synthesized via exfoliation of graphite oxide in hydrogen environment at low temperature while graphite oxide (GO) was prepared normally by Hummers method.

2.2. Structure, electrical, chemical composition and surface area characterization

X-ray diffraction (XRD), scanning electron microscopy/energy-dispersive analysis (SEM/EDX), high-resolution transmission electron microscopy (HRTEM), infrared spectra (IR) and the Brunauer–Emmett–Teller (BET) specific surface areas were the most common analytical techniques to characterize the morphologies, structures, chemical composition and surface area of nanocomposite electroactive materials. XRD analysis was carried out for the nanocomposite samples containing metal oxides to examine the crystallinity and crystal phases of the oxide materials. IR spectra were used for the identification of the characteristic bands of a polymer for nanocomposites consisting of conducting polymers. SEM was used for morphological analysis of nanocomposites, while EDX was used to determine their chemical composition. The electrical conductivity of nanocomposites was obtained normally using a four-probe technique. To do the measurement, the nanocomposite samples were ground into fine powders and then were pressed as pellets. The weight loss of nanocomposite material and the heat flow associated with the thermal decomposition during synthesis or heat treatment were studied by thermogravimetric analysis (TGA) and differential thermal analysis (DTA).

2.3. Electrochemical characterizations

Cyclic voltammetry (CV) was usually conducted to characterize the nanocomposite electrode in a three-electrode cell in either aqueous electrolytes or organic electrolytes using an electrochemistry workstation. The working electrode was metal plate or mesh (e.g. nickel, aluminium, stainless steel) coated with a mixture of nanocomposite and conductive carbon such as acetylene black with a binder such as PTFE or polyvinylidenedifluoride (PVDF). The reference electrodes were either saturated calomel electrode (SCE), Ag/AgCl or others. The counter electrode was typically platinum foil. The specific current and specific capacitance of nanocomposite was determined by the CV current value, scan rate and the weight of nanocomposites.

Galvanostatic charge-discharge cycling was performed with two electrode system having identical electrodes made of same nanocomposite electroactive material. Constant current densities ranging from 0.5 to 10 mA/cm^2 were typically employed for charging/discharging the cell in the voltage range 0-1 V for aqueous electrolytes or 0-2.7 V for organic electrolytes. The discharge capacitance (C) is estimated from the slope (dV/dt) of the linear portion of the discharge curve using the expression.

$$C = I \, dV / dt \tag{1}$$

The weight of the active material of the two electrodes is same in a symmetric supercapacitor. The specific capacitance (Cs) of the single electrode can thus be expressed as:

$$C_s = 2C / m \tag{2}$$

where m is the active material mass of the single electrode. The energy density (E_d) of the capacitor can be expressed as,

$$E_d = \frac{1}{2} \left(C_s V^2_{max} \right) \tag{3}$$

The coulomb efficiency 'η' was evaluated using the following relation,

$$h = t_d / t_c \times 100\% \tag{4}$$

where t_c and t_d are the time of charge and discharge respectively.

Experiments of electrochemical impedance spectra (EIS) were also performed with a two electrode system having identical electrodes made of same nanocomposite active electrode materials at open circuit potential (OCP) over the frequency range 10 kHz–10mHz with a potential amplitude normally of 5 mV. The impedance spectra usually show a single semi-circle in the high frequency region and nearly vertical line in the low frequency region for a supercapacitor, which indicates that the electrode process is controlled by electrochemical reaction at high frequencies and by mass transfer at low frequencies. The intercept of the semi-circle with real axis (Z_{real}) at high frequencies is the measure of internal resistance (R_s) which may be due to (i) ionic resistance of the solution or electrolyte, (ii) intrinsic resistance of the active electrode materials and (iii) interfacial resistance between the electrode and current collector. The origin of the semi-circle at higher frequency range is due to ionic charge transfer resistance (R_{ct}) at the electrode–electrolyte interface. The diameter of the semi-circle along the real axis (Z_{dia}) gives the charge transfer resistance R_{ct}.

3. Performance of various types of nanocomposite active electrode materials

3.1. Mixed pseudocapacitive metal oxide nanocomposites

Metal oxides like such as RuO_2, MnO_2, Co_3O_4, NiO, SnO_2, Fe_3O_4, and V_2O_5 have been employed as electroactive materials for pseudocapacitors. Those metal oxides typically have several redox states or structures and contribute to the charge storage in pseudocapacitors via fast redox reactions. The remarkable performance of RuO_2 in supercapacitors (exhibits the highest specific capacitance values of 720 F g^{-1}) has stimulated many interests in investigating metal oxide system for supercapacitor applications. The commercial use of RuO_2, however, is limited owing to its high cost and toxic nature. Other simple metal oxides usually have some limitations such as poor electrical conduction, insufficient electrochemical cycling stability, limited voltage operating window and low specific capacitance. Those limitations need to be addressed in order for commercial applications of supercapacitors based on metal oxides. Mixed binary or ternary metal oxides systems, such as Ni–Mn oxide, Mn–Co oxide, Mn–Fe oxide, Ni–Ti oxide, Sn–Al oxide, Mn–Ni–Co oxide, Co–Ni–Cu oxide and Mn–Ni–Cu oxide have shown improved properties as electroactive materials for pseudocapacitors and have shed new lights in this area of research. The following section summarizes the recent development in seeking electroactive mixed metal oxide nanocomposites for pseudocapacitors.

3.1.1. Mixed manganese oxides

The natural abundance and low cost of Mn oxides, along with their satisfactory energy-storage performance in mild electrolytes and environmental compatibility, has made them the most promising new electroactive material for the pseudocapacitor applications. However, Mn oxides has limitations such as low surface areas, poor electrical conductivity and relatively small specific capacitance value. To improve the electrochemical performance of Mn oxides for pseudocapacitors, many efforts have been devoted to incorporate various transition metals into Mn oxides to form mixed metal oxide nanocomposites with controlled micro-/nanostructures in order to improve their electrochemical characteristics. The understanding of their synergistic effect and the eventual design of an integrated material architecture in which each component's properties can be optimized and a fast ion and electron transfer will be guaranteed still remains a great challenge.

Jiang et al. [1] designed and synthesized MnO_2 nanoflakes-$Ni(OH)_2$ nanowires composites that can be used in both neutral and alkaline electrolytes and have very high cycling stability. The nanocomposites with 70.4 wt.% MnO_2 content exhibited specific capacitance of 355 F g^{-1} with excellent cycling stability (97.1% retention after 3000 cycles) in 1 M Na_2SO_4 neutral aqueous solution. In 1M KOH aqueous alkaline solution, the MnO_2-$Ni(OH)_2$ nanocomposite with 35.5 wt. % MnO_2 content possessed a specific capacitance of 487.4 F g^{-1} also with excellent cycling stability. Such excellent capacitive behaviours are attributed by the authors to the unique MnO_2–$Ni(OH)_2$ core–shell nanostructures as depicting in Figure 1(b). The interconnected MnO_2 nanoflakes were well-dispersed on the surface of $Ni(OH)_2$ nanowires that creates highly porous surface morphology. This integrated structure can provide high surface area and more active sites for the redox reactions. The specific capacitance and Coulombic efficiency as function of cycle number at a current density of 10 A g^{-1} for up to 3000 cycles is also shown in Figure 1(a) for the MnO_2-$Ni(OH)_2$ nanocomposite. After long cycling, the $Ni(OH)_2$–MnO_2 nanocomposites are overall preserved with little structural deformation, as shown in Figure 1(c) and (d). Oxides of Pb, Fe, Mo, and Co were also incorporated into MnO_2 to form mixed metal oxide nanocomposites. Kim et al.[2] synthesized mixed oxides of Mn with Pb or Ni by reduction of $KMnO_4$ with either lead(II) acetate-manganese acetate or nickel(II) acetate-manganese acetate reducing solutions. Characterization of the nanocomposite electrodes were carried out using cyclic voltammetry, galvanostatic charge-discharge, XRD, BET analysis, and TGA. The results showed that by introducing Ni and Pb into MnO_2, the surface area of the mixed oxide increased due to the formation of micropores. The specific capacitance increased from 166 F g^{-1} (for MnO_2) to 210 and 185 F g^{-1} for Mn-Ni and Mn-Pb mixed oxides, respectively. Kim et al. [2] also found that annealing of the nanocomposites can affect their capacitance: transition from amorphous to a crystalline structure occurred at high temperature (400 °C) reduces the specific capacitance. Binary Mn–Fe oxide was electroplated on graphite substrates by Lee et al. [3] at a constant applied potential of 0.8V vs. SCE in a mixed plating solution of $Mn(CH_3COO)_2$ and $FeCl_3$. The electrochemical behaviours of the as-deposited and the annealed mixed oxide nanocomposites were characterized by cyclic voltammetry in 2M KCl solution. Lee et al. found that as-deposited Mn–Fe binary oxide has porous structure and is amorphous. After annealing at 100°C to remove the adsorbed water, the partially hydrous mixed oxide has optimized ionic and electronic conductivity and gives rise to the best pseudocapacitive performance. However, if the annealing temperature is increased to higher, the mixed oxide loses it porosity and slowly crystalizes which leads to the decrease in specific capacitance. A series of Mn and Mo mixed oxides (i.e. Mn-Mo-X (X= W, Fe, Co)) were investigated by Ye et al.[4] and they found that the specific capacitance of Mo doped Mn oxides are higher than that of pure Mn oxide. The Mn-Mo-Fe oxide reach a high specific capacitance value of 278 F g^{-1} in aqueous 0.1 M Na$_2$SO$_4$ electrolyte at a scan rate of 20 mV s^{-1} and has a rectangular-shaped voltammogram. The improvement in capacitance of Mn oxides doped with molybdenum was attributed by the authors to the formation of nanostructure and the existence of low crystallinity. The above results show that mixed metal oxides with amorphous structure have better specific capacitance than that of crystalline structure. Incorporation of various transition metals into Mn oxides creates more porous structures, therefore increase their specific capacitance.

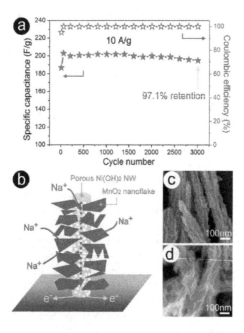

Figure 1. a) Specific capacitance as a function of cycle number at 10 A g^{-1}, (b) schematic of the charge storage advantage of the Ni(OH)$_2$–MnO$_2$ core–shell nanowires, (c) and (d) SEM images of the Ni(OH)$_2$–MnO$_2$ core–shell nanowires before and after 3000 cycles (from ref. 1)

Advance thin film physical vapour deposition methods were also used to prepare mixed metal oxide nanocomposite for supercapacitor electroactive materials research. Thin films of manganese oxide doped with various percentages of cobalt oxide were grown by pulsed laser deposition (PLD) on silicon wafers and stainless steel substrates at our laboratory [5]. Before investigated Co-doped manganese oxide film, our team [6] developed different PLD processing parameters (i.e. temperature, oxygen pressure) to produce various chemical compositions and phases of manganese oxides such as pure crystalline phases of Mn_2O_3 and Mn_3O_4 as well as amorphous phase of MnO_x. He then evaluated the pseudo-capacitance behaviours of these different phases of manganese oxides and found that the crystalline Mn_2O_3 phase has the highest specific current and capacitance, while the values for crystalline Mn_3O_4 films are the lowest. The specific current and capacitance values of the amorphous MnO_x films are in between Mn_2O_3 and Mn_3O_4. The specific capacitance of Mn_2O_3 films of 120 nm thick reaches 210 F g^{-1} at 1 mV s^{-1} scan rate with excellent stability and cyclic durability. He then doped amorphous MnO_x and crystalline Mn_2O_3 phases with Co_3O_4 and characterized the mixed Co-Mn oxide films with X-ray diffraction and CVs. The CVs recorded at a 20 mV s^{-1} scan rate for un-doped and Co-doped amorphous MnO_x films are shown in Figure 2(a), and their specific capacitance determined from the CV curves at scan rates of 5, 10, 20 and 50 mV s^{-1} are shown in Figure 2(b). The CVs in Figure 2 shows that the Co-doped amorphous MnO_x films have larger specific currents and capacitances than the un-doped amorphous MnO_x film. Low cobalt doping (3.0 atm.%) had the greatest increase in capacitance, followed by 9.3 atm.% cobalt doping. The 22.6 atm.% cobalt doping had the least increase in specific capacitance. The operating potential window (between H_2 evolution and O_2 evolution due to decomposition of water) was shifted about 100 mV toward more negative potentials for all the Co-Mn mixed oxide films. At a 5 mV/s scan rate, the 3.0 atm.% Co-doped MnOx film reached 99 F g^{-1}, which is more than double that the 47 F g^{-1} observed for the un-doped MnO_x film. This result indicates that Co doping significantly improves the pseudo-capacitance performance of amorphous manganese oxide. However, Co-doped crystalline Mn_2O_3 films did not show an improvement in specific current and capacitance compared with un-doped Mn_2O_3 crystalline films. High Co doping level (20.7 atm.% doped) in the crystalline Mn_2O_3 films actually decreased both the specific current and capacitance values. These findings demonstrate that elemental doping is an effective way to alter the performance of pseudo-capacitive metal oxides. Our work also demonstrated that thin film deposition techniques such as PLD are very promising techniques for screening high performance mixed oxide active materials for supercapacitor applications.

3.1.2. Other mixed metal oxides

In addition to mixed manganese oxides, many other mixed metal oxides have also been investigated as electroactive materials for supercapacitor applications. Co_3O_4-$Ni(OH)_2$ nanocomposites were synthesized by electrochemical deposition on the Ti substrate in a solution of $Ni(NO_3)_2$, $Co(NO_3)_2$ and NH_4Cl, then follows by heat treatment at 200ºC [7]. The Co_3O_4-$Ni(OH)_2$ electrodes exhibited high specific capacitance value of 1144 F g^{-1} at 5 mV s^{-1} and long-term cycliability. The excellent capacitive behaviours of Co_3O_4-$Ni(OH)_2$ nanocomposite was attributed by the authors to the porous network structures that favour electron and ions transportation as well as faradic redox reactions of both couples of Co^{2+}/Co^{3+} and Ni^{2+}/Ni^{3+}. Y. Yang et al. [8] prepared mixed V_2O_5–TiO_2 nanotube arrays by anodizing Ti–V alloys with different V compositions using an ethylene glycol with 0.2 M HF as the electrolyte at a comparably high anodization voltage. Well-defined nanotube structures were grown for alloys with vanadium content up to 18 at%. The mixed V_2O_5–TiO_2 nanotube arrays were found to exhibit greatly enhanced capacitive properties compared with pure TiO_2 nanotubes. The specific capacitance of the mixed V_2O_5–TiO_2 nanotubes can reach up to 220 F g^{-1} with an energy density of 19.56 Wh kg^{-1} and was found to be very stable in repeated cycles. Another interesting mixed oxide is SnO_2–Al_2O_3 mixed oxide [9], which shows much greater electrochemical capacitance than pure SnO_2 and was electrochemically and chemically stable even after cycling1000 times.

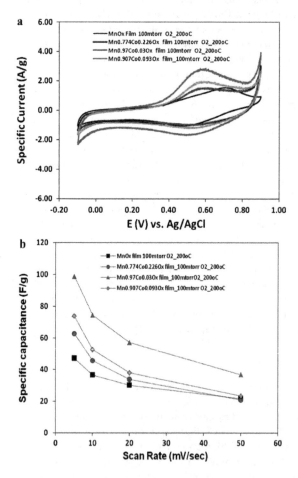

Figure 2. Cyclic voltammetry (a) and specific capacitance (b) of amorphous MnO$_x$ film and various Co-doped amorphous MnO$_x$ films deposited by PLD at 200∘C in 100 mTorr of O$_2$ (from ref. 6).

Spinel nickel cobaltite (doped or un-doped, such as $NiCo_2O_4$ and $NiMn_xCu_{2-x}O_{4-y}$ ($x\leq1.0$)) possesses much better electronic conductivity than that of NiO and Co_3O_4. They are low-cost and have multiple oxidation states, and therefore, are also exploited for supercapacitor applications. C. Wang et al. [10] prepared nanostructured $NiCo_2O_4$ spinel platelet like particles with narrow size distribution of 5–10 nm by co-precipitation process. The $NiCo_2O_4$ has excellent conductivity and showed a high-specific capacitance of 671 F g^{-1} under a mass loading of 0.6 mg cm^{-1} at a current density of 1 A g^{-1}. Chang et al. [11] also prepared $NiCo_2O_4$ and $NiMn_xCo_{2-x}O_{4-y}$ (x≤1.0) using a precipitation route. They found that the spinel structural of $NiCo_2O_4$ is retained with a quarter of the Co ions replaced with Mn. The presence of Mn significantly suppresses crystallite growth upon thermal treatment, and greatly enhances the specific capacitance of the spinel. At the scan rate of 4 mV s^{-1}, the specific capacitance is found to increase from 30 F g^{-1} for Mn content x = 0 to 110 F g^{-1} for x = 0.5. The $NiMn_{0.5}Co_{1.5}O_4$ powder has been found by the authors to be much smaller surface area than the $NiCo_2O_4$ powder. Therefore, the remarkable capacitance enhancement exhibited by the $NiMn_{0.5}Co_{1.5}O_4$ electrode is not due to microstructural variations of the oxide powders. The capacitance enhancement is attributed by the authors to the facile charge-transfer characteristic of the Mn ions, which enables a greater amount of charge transferred between the oxide and the aqueous electrolyte species over the same potential window.

3.2. Carbon nanotubes based nanocomposites

Carbon nanotubes (CNTs) have superior material properties such as high chemical stability, aspect ratio, mechanical strength and activated surface area as well as outstanding electrical properties, which make them good electroactive material candidates for supercapacitors. The electrodes made from CNTs exhibit a unique pore structure for change storage; however, there are limitations for further increasing the effective surface area of the CNTs, as well as relatively high materials cost which limit the commercial application of CNTs based supercapacitors. To improve the performance of CNTs, they are composited with conductive polymers and metal oxides. This section will summarize the recent development of CNTs based nanocomposites for supercapacitor applications.

3.2.1. Carbon nanotubes and polymer nanocomposites

Techniques that can be used to synthesize CNTs include Arc discharge, chemical vapour deposition, and laser ablation. Kay et al. [12] synthesized single-walled CNTs by dc arc discharge of a graphite rod under helium gas using Ni, Co, and FeS as catalysts. Then they prepared single-walled CNTs-polypyrrole (PPY) nanocomposite using in situ chemical polymerization of pyrrole monomer in solution with single-walled CNTs suspension. Figure 3 shows the FE-SEM images of as-grown single-walled CNTs, pure PPY, and single-walled CNT-PPY nanocomposite powder formed by the in situ chemical polymerization. The as-grown single-walled CNTs are randomly entangled and cross-linked, and some carbon nanoparticles are also observed, as shown in Figure 3(a). In Figure 3(b), the image for pure PPY synthesized without single-walled CNTs present in solution shows a typical granular morphology with granule size of about 0.2-0.3 mm. Figure 3(c) demonstrates that the individual carbon nanotube bundles are uniformly coated with PPY which indicates that in situ chemical polymerization of pyrrole can effectively coated all the CNTs. The electrode prepared using single-walled CNTs-PPY nanocomposite as active materials show very high specific capacitance: a maximum specific capacitance of 265 F g^{-1} from the single-walled CNT-PPY nanocomposite electrode containing 15 wt. % of the conducting agent was obtained. Figure 4 shows the specific capacitances of the as-grown single-walled CNTs, pure PPY, and single-walled CNTs-PPY nanocomposite electrodes as a function of discharge current density. In comparison to the pure PPY and as-grown single-walled CNTs electrodes, the single-walled CNTs-PPY nanocomposite electrode shows very high specific capacitance. The improvement in the specific capacitance of the CNTs-PPY nanocomposite was attributed by the authors to the increase in active surface area of pseudocapacitive PPY by CNTs. CNTs was also composited with polyaniline (PANI) by Deng et al. [13]. In their experiments, the CNTs–PANI was prepared using direct polymerization of aniline monomer with oxidant agent, $(NH_4)_2S_2O_8$, in acidic solution containing CNTs suspension, similar to the CNTs-PPY nanocomposite prepared by Kay et al. The CNTs–PANI nanocomposite achieved a specific capacitance of 183 F g^{-1}, almost 4 times higher than pure CNTs (47 F g^{-1}).

Figure 3. The FE-SEM images of (a) as-grown single-walled CNTs, (b) pure PPY, and (c) single-walled CNTs-PPY powder (from ref. 12)

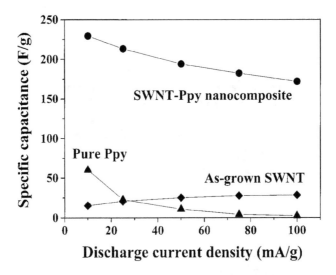

Figure 4. The specific capacitances of the as-grown single-walled CNTs, pure PPY, and single-walled CNTs-PPY, single-walled CNTs-PPY nanocomposite electrodes as a function of ischarge current density at a charging voltage of 0.9 V for 10 min (from ref.

3.2.2. Carbon nanotubes and ruthenium oxide nanocomposites

Although RuO_2 shows remarkable performance as supercapacitors electrode active materials, its high cost has limited its commercial applications. To fully utilize the expensive RuO_2 in an electrode, it is necessary to disperse it over high surface area materials such as CNTs. Y-T. Kim et al. [14, 15] discovered a new way to uniformly disperse RuO_2 over the whole surface area of CNTs: firstly, they oxidized the multi-walled CNTs in a concentrated H_2SO_4–HNO_3 mixture to introduce the surface carboxyl groups, and then they prepared multi-walled CNTs-RuO_2 nanocomposites with a conventional sol–gel method. The surface carboxyl groups formed on CNTs allow RuO_2 to disperse more effectively since bond formation between RuO_2 and carboxyl group protects against agglomeration of RuO_2 as illustrated in Figure 5. Figure 5 schematically shows that for purified multi-walled CNTs, RuO_2 can be spontaneously reduced to metallic Ru on the CNTs surface and subsequently covered with $RuO_x(OH)_y$ via the reaction with NaOH to form core-shell structures. For oxidized CNTs, the positive charged Ru precursor ions have limited contact with the CNTs surface to be reduced due to negatively charged carboxyl groups. Surface carboxyl groups act not only as protectors against spontaneous reduction of Ru ions but also as anchorage centres for Ru which enhance the dispersity of RuO_2 and hinder their agglomeration into large particles. TEM images in figure 5 show the dramatic difference in particle size of RuO_2 nanoparticle and dispersity between RuO_2 –pure CNTs and RuO_2–oxidized CNTs.

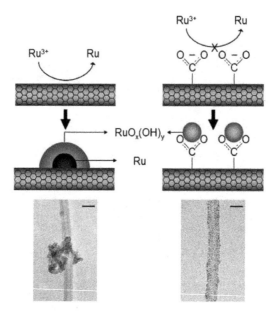

Figure 5. Schematic diagram of the different formation mechanisms of RuO_2 on purified multi-walled CNTs and oxidized multi-walled CNTs in the preparation process and their actual TEM images scale bar (20 nm). (from ref. 15)

Another way to effectively disperse RuO_2 over CNTs was developed by Hsieh et al. [16] who grew vertically aligned multi-walled CNTs films directly by CVD on the Ti current collector using thin nickel layers as the catalyst. Hydrous ruthenium dioxide was then directly deposited onto the surface of CNTs electrodes by electrochemical CV deposition from an aqueous acidic solution of ruthenium trichloride ($RuCl_3.nH_2O$). The SEM morphology of the composites shows that the surfaces of the multi-walled CNT/Ti electrodes were coated uniformly with hydrous ruthenium dioxide, which increased the utilization of the electroactive RuO_2 material. Electrochemical measurements showed that the $RuO_2.nH_2O$-CNTs nanocomposites have high capacitance of 1652 F g^{-1} and decay rate of 3.45% at 10 mV s^{-1} in a 1.0 M H_2SO_4 aqueous electrolyte within the potential range from -0.1 to 1.0 V. Figure 6 shows the comparison of specific capacitance of $RuO_2.nH_2O$-CNTs nanocomposite electrode with $RuO_2.nH_2O$ and multi-walled CNTs electrodes at a scan rate of 10 mV/s. The results in the figure 6 clearly show that the capacitance of $RuO_2.nH_2O$-CNTs was much higher (4.7 times) than those of pure materials. Chemical impregnation of ruthenium nitrosylnitrate solution ($Ru(NO)(NO_3)_x(OH)_y$) on the CVD-grown multi-walled CNTs following by calcination at 350°C process was used by Lee et al. [17] to form RuO_2-CNTs nanocomposites. The specific capacitance of the nanocomposite was found to be as high as 628 F g^{-1}. The authors believed that nanoporous three-dimensional structure of RuO_2-CNTs nanocomposite facilitated the electron and ion transfer. Byung Chul Kim et al. [18] used the electrochemical potentiodynamic deposition method to prepare RuO_2-CNTS and Ru/Co oxides-CNTs nanocomposites from $RuCl_3$, and 0.1M $CoCl_2$ + 0.05M $RuCl_3$ solutions, respectively. All the composites showed considerable increase in capacitance values. The Ru/Co mixed oxides-CNTs showed superior performance (570 F g^{-1}) at high scan rates (500 mV s^{-1}) when compared to the RuO_2 electrode (475 F g^{-1}). This increase in capacitance at high scan rates is attributed by the authors to the enhanced electronic conduction of Co in the composites.

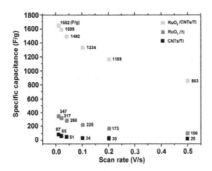

Figure 6. Specific capacitance of $RuO_2.nH_2O$-CNT, $RuO_2.nH_2O$, and multi-walled CNTs after 80 scan cycles with scan rates from 10 to 500 mV/s (from ref. 16).

3.2.3. Carbon nanotubes and other metal oxide nanocomposites

Besides RuO_2, there are many other metal oxides that were composited with CNTs to form electroactive materials for supercapacitors. Chen et al. [19] used thermal CVD to grow multi-walled CNTs on a stainless steel plate and then on top of CNTs, IrO_2 nanotubes were deposited using metal-organic CVD with the iridium source of $(C_6H_7)(C_8H_{12})Ir$ at 350∘C. The IrO_2 square nanotube crystals were grown on the upper section of the CNTs thin film. Figure 7 shows the morphologies of multi-walled CNTs, IrO_2 nanotubes, and IrO_2 nanotubes – multi-walled CNTs nanocomposite. The cross-sectional view of multi-walled CNTs, figure 7(b), shows that there are upper and lower sections in the CNTs thin film. The upper section, ~2 μm in thickness, consists of entangled carbon nanotubes without distinct orientation. The lower section, approximately 4 μm thick, is composed of largely parallel nanotubes, aligned in the vertical direction. These nanotubes act as the templates for the IrO_2 nanotube growth. Figures 7(c) and (d) show a top view and cross-sectional view of IrO_2 nanotubes grown on stainless steel substrate. Figure 7(e) and (f) show IrO_2 nanotubes grown over CNTs. Figure 7(e) indicates that the grown IrO_2 nanotubes had a high density along the wires of multi-walled CNTs in the upper section. In comparison to multi-walled CNTs, the nanostructured IrO_2–CNTs increases the capacitance by a factor of six, from 15 to 69 F g^{-1}, and reduces the resistance from 90 to 60 Ω. Such a hierarchical structure provides a high surface area for electrical charge storage, and a double-layer capacitance in conjunction with pseudocapacitance. Electrochemical anodic deposition of $MnO_x \bullet nH_2O$ films from $MnSO_4 \bullet 5H_2O$ solution on CNTs coated Ni substrates was used by Lee et al. [20] to form amorphous MnO_x-CNTs nanocomposite electrode. The CNTs were electrophoretically deposited on the Ni substrate by applying a dc voltage of 20V before the deposition of $MnOx \bullet nH2O$. The MnO_x-CNTs nanocomposite electrodes have shown much better energy storage capabilities than MnO_x deposited directly on the Ni substrate: the specific capacitances were 415 as obtained from CV measurements with a scan rate of 5 mV/s and it preserved 79% of its original capacitance value after 1000 cycles. The authors attributed the improvement to the low resistance and large surface area of the nanocomposite electrodes. Other metal oxide CNTs nanocomposites being investigated include NiO-CNTs[21], V_2O_5-CNTs [22] and SnO_2-V_2O_5-CNTs [23]. The V_2O_5-CNTs composite was prepared by electrochemical deposition of V_2O_5 on vertically aligned multi-walled CNTs and it can reach a specific capacitance of 713.3 F g^{-1} at 10 mV s^{-1}. The SnO_2-V_2O_5-CNTs was prepared by simply mixing CNTs and SnO_2-V_2O_5 mixed oxide powder in a mortar prior and fixing on the surface of a graphite electrode that was impregnated with paraffin. At a scan rate of 100 mV s^{-1}, the SnO_2-V_2O_5-CNTs electrode provides 121.4 F g^{-1} specific capacitance.

Other types of high surface area carbon materials such as activated carbons, carbon fibres and carbon aerogels were also composited with metal oxide to form high performance electroactive materials. Those examples include vapour-grown carbon fibre (VGCF) - RuO_2 xH_2O nanocomposite prepared by a thermal decomposition [24], $RuO_2.xH_2O$-mesoporous carbon nanocomposites prepared using impregnation [25], ZnO–activated carbon nanocomposite electrode by simply mixing [26] and MoO_3-graphite prepared by ball milling [27].

3.3. Pseudocapacitive polymer and metal oxide nanocomposites

Electronically conducting polymers derived from monomers such as pyrrole, aniline, and thiophene have unique properties, such as good environmental stability, electroactivity, and unusual doping/de-doping chemistry, therefore, they are suitable for active electrode material usage in supercapacitors. When used as electrode materials, these polymers have advantage over carbon-based materials since they have both electrochemical double layer capacitance and pseudocapacitance which arises mainly from the fast and reversible oxidation and reduction processes related to the π-conjugated polymer chain. However, conducting polymers have problems of typical volumetric shrinkage during ejection of ions (doped ions) and low conductance at de-doped state which would result in high ohmic polarization of supercapacitors. In order to solve this problem, conducting polymers were mixed with metal oxides to form nanocomposites and the synergistic effect of the polymer–metal oxide nanocomposites has been exploited. Such nanocomposites were found to have the advantages of polymers such as flexibility, toughness and coatability and metal oxides such as hardness, and durability. They also possess some synergetic properties which are different from that of parent materials. This section will summarize the recent development in this serial of materials.

Figure 7. SEM image of CNTs top view (a), CNTs cross-sectional view (b), IrO_2 nanotubes top view (c), IrO_2 cross-sectional view (d), IrO_2–CNTs top view (e), IrO_2–CNTs cross-sectional view (f). The inset of (b) and (d) are magnified images. (from ref. 19).

3.3.1. Polyaniline (PANI) and metal oxide nanocomposites

Polyaniline (PANI) is one of the most important conducting polymers because of its ease of synthesis at low cost, good processability, environmental stability and easily tuneable conducting properties. The synthesis and studies of composites of PANI and metal oxides such as MnO_2, SnO_2 and $MnWO_4$ have been carried out. In a PANI-metal oxide nanocomposite, PANI not only serves as an electroactive material for energy storage but also as a good coating layer to restrain metal oxides from dissolution in acidic electrolytes. Chen et al. [28] synthesized a very high performance PANI-MnO_2 nanocomposite using the following procedure: first, the hydroxylated MnO_2 nanoparticles were surface modified with silane coupling agent, ND42, then the obtained surface modified MnO_2 nanoparticles (ND-MnO_2) were washed and dried. Electro-co-polymerization of aniline and ND-MnO_2 nanoparticles was conducted on a carbon cloth in an electrolyte solution containing ND-MnO_2, aniline, H_2SO_4 and Na_3PO_3. The co-polymerization was preceded through successive cyclic voltammetric scans. The whole synthesis process is illustrated in Figure 8. Electro-co-polymerization method was also used to prepare unmodified PANI–MnO_2 nanocomposite and pure

PANI. The SEM images of PANI–MnO$_2$, PANI-ND-MnO$_2$ films reveal that the addition of MnO$_2$ nanoparticles promotes the one dimensional growth of PANI, which substantially reduces the size of the nanorods and increases the surface area/internal space of the composite films (Figure 9). PANI-ND-MnO$_2$ composite film has an average specific capacitance of ~80 F g^{-1} and a very stable coulombic efficiency of ~98% over 1000 cycles. It also exhibit high intrinsic electrical conductivity and good kinetic reversibility. The excellent properties were attributed by the authors to the improved interaction between MnO$_2$ and PANI and the increased effective surface area in PANI-ND-MnO$_2$ film, due to the surface modification of MnO$_2$ nanoparticles with the silane coupling reagent. Significantly high specific capacitor was achieved with PANI-SnO$_2$ nanocomposites prepared by Hue et al. [29] using a chemical method in which SnO$_2$ nanoparticles and aniline were dispersed in sodium dodecylbenzene-sulfonate solution and then, ammonium persulfate was added to the above mixture to start polymerization. The PANI-SnO$_2$ nanocomposite thus prepared had a high specific capacitance of 305.3 F g^{-1} with a specific energy density of 42.4Wh kg^{-1} and a coulombic efficiency of 96%. The energy storage density of the composite was about three times as compared with pure SnO$_2$.

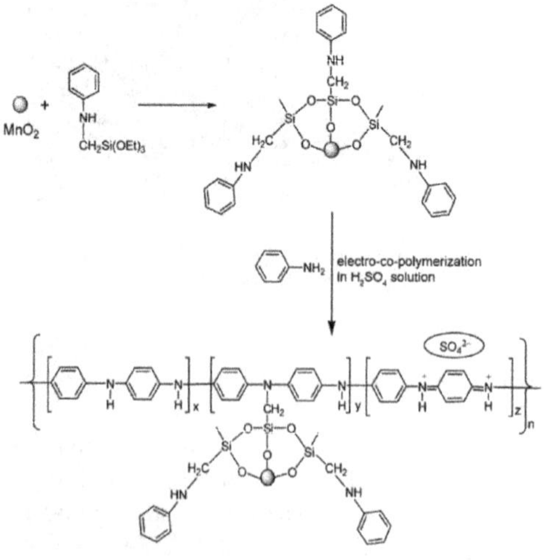

Figure 8. A schematic diagram illustrates the reaction pathway for the synthesis of PANI-ND-MnO$_2$ nanocomposite film (from ref. 28).

Wang et al. [30] developed an innovative way to synthesize PANI-MnO$_2$ nanocomposites. This so-called "interfacial synthesis" utilized the interfacial region between an organic phase and an aqueous phase to synthesize the composite. The organic phase was prepared by dissolving aniline monomers into inorganic Trichloromethane (CHCl$_3$) solution, while the aqueous phase was obtained by dissolving potassium permanganate in distilled water. When the aqueous solution was added into the organic solution, an interface was formed immediately between the two phases and the reaction occurred. During the reaction, aniline was diffused from the organic solution to the interface and was chemically oxidized into polyaniline. At the same time, MnO_4^- was reduced to manganese oxide precipitate. Finally, the PANI-MnO_2 nanocomposite was formed and remained in the aqueous solution. For comparison, conventional chemical co-precipitation of MnO_2-PANI composite was performed to make the conventional PANI-MnO_2 composite. Both synthesis processes was schematically illustrated in Figure 10. The interfacial synthesized MnO_2-PANI composite shows larger specific surface area (124 m^2 g^{-1}) and more uniform pore-size distribution than the composite prepared by chemical co-precipitation as shown in Figure 11. It exhibits a higher specific capacitance of 262 F g^{-1} (about twice amount of conventionally prepared MnO_2-PANI composite) with better cycling stability. The authors attributed the observed enhanced electrochemical properties of the interfacial synthesized MnO_2-PANI composite electrode to its unique hollow microstructure with well-defined mesoporosity and the coexistence of conducting PANI. Other interesting PANI metal oxide nanocomposites include PANI-$MnWO_4$ nanocomposites, which was prepared in situ polymerization of aniline monomer in solution containing $MnWO_4$ nanoparticles [31]. The composite has shown good electrochemical properties: with 50% of $MnWO_4$ loading, the PANI-$MnWO_4$ nanocomposites shows high specific capacitance of 475 F g^{-1}, much higher than that of the physical mixture of PANI and $MnWO_4$ (346 F g^{-1}).

3.3.2. Other polymer and metal oxide nanocomposites

Beside Polyaniline (PANI), other conducting polymers that have been used to composite with metal oxides to form electroactive materials including polypyrrole (PPY), polythiophene and their derivatives such as Poly 3,4-ethylenedioxythiophene (PEDOT). PEDOT is a stable and environmentally friendly polymer and has controllable electrical conductivity. However, PEDOT suffers from problems such as volumetric swelling and shrinkage during the insertion and ejection of ions. PEDOT was comprised with pseudocapacitive metal oxides such as $MnFe_2O_4$, $CoFe_2O_4$ and MoO_3 to improve its property. The synergistic effect of composite formation plays a significant role to increase the capacitance value. Sen et al. [32] prepared PEDOT–$NiFe_2O_4$ nanocomposites by chemical polymerization of EDOT monomer in solution containing nickel ferrite nanoparticles ($NiFe_2O_4$). Pure PEDOT polymer in both n-hexane medium and aqueous medium was also synthesized by similar procedure in the absence of $NiFe_2O_4$ nanoparticles. Electrochemical CVs and impedance spectroscopy were used to characterize the PEDOT–$NiFe_2O_4$ nanocomposite as well as pure PEDOT synthesized in organic medium and aqueous medium, and $NiFe_2O_4$ nanoparticles prepared by sol–gel procedure. Figure 12 shows typical Nyquist impedance spectra of the four compounds over a frequency range of 10 kHz–10 mHz with a potential amplitude of 5mV. The impedance results show that introduction of $NiFe_2O_4$ nanoparticles into the PEDOT not only helps to reduce the intrinsic resistance (the intercept of the semi-circle with real axis (Z′) at high frequencies is the measure of internal resistance) through the development more mesoporous structures but also increase the kinetics of electron transfer through redox process leading to the enhancement of pseudocapacitance in the composite materials (pseudocapacitance values were also determined from the impedance by fittings the spectra with Randles equivalent circuit). The PEDOT– $NiFe_2O_4$ nanocomposite shows high specific capacitance (251 F g^{-1}) in comparison to $NiFe_2O_4$ (127 F g^{-1})

and PEDOT (156 F g^{-1}) where morphology of the pore structure was believed to play a signifi-cant role over the total surface area. PEDOT was also composited with MoO$_3$ by Murugan et al. [33] using chemical polymerization of EDOT monomer with FeCl$_3$ as oxidizing agent in MoO$_3$ suspension. The nanocomposite also has much higher specific capacitance (300 F g^{-1}) com-pared to that of pristine MoO$_3$ (40 mF g^{-1}). The improved electrochemical performance was at-tributed by the authors to the intercalation of electronically conducting PEDOT between MoO$_3$ layers and an increase in surface area.

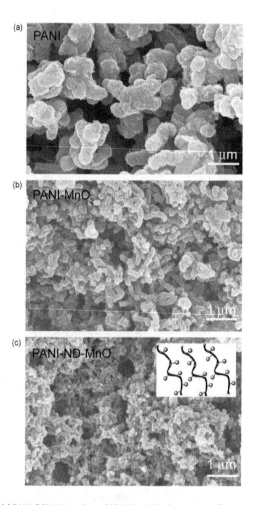

Figure 9. SEM images of (a) PANI, (b)PANI–MnO$_2$, and (c)PANI-ND-MnO$_2$ composite(from ref. 28).

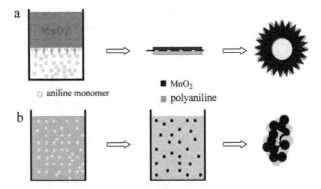

Figure 10. Schematic illustration of the formation mechanisms of MnO_2-PANI composites: (a) interfacial synthesis and (b) chemical co-precipitation (from ref. 30).

Polypyrrole (PPY) is also a promising conducting polymer material due to its highly reversible redox reaction. Although the electrical conductivity of intrinsic PPY is low, doping of surfactants can enhance effectively the electrical conductivity of PPY. p-Toluenesulfonic acid (p-TSA) was used as a dopant by Dong et al. [34] to prepare MnO_2-PPY/TSA nanocomposite for supercapacitor applications. TSA and pyrrole were dispersed ultrasonically in deionised water to form a homogeneous solution. With the addition of $KMnO_4$ or $FeCl3 \bullet 6H2O$ oxidant, redox reactions occurred and MnO_2-PPY/TSA nanocomposite was produced. Micrographs and BET isotherm measurements showed that the particle and the pore size of the MnO_2-PPY/TSA nanocomposite are much smaller than those of the MnO_2-PPY. Electrochemical measurements showed that the MnO_2-PPY/TSA nanocomposite electrode exhibited a higher specific capacitance of ~376 F g^{-1} at 3 mA cm^{-2} and better cycling stability in 0.5M Na_2SO_4 solution than the MnO_2-PPY. Another polymer metal oxide composite that shows promising supercapactive properties is MnO_2-poly(aniline-co-o-anisidine) [35], which has specific capacitance of the 262 F g^{-1} in 1M Na_2SO_4 at a current density of 1A g^{-1}. All the above results presented by various authors have demonstrated that the development of novel metal oxide-conducting polymers nanocomposite holds great potential applications in high-performance electrochemical capacitors.

3.4. Graphene based nanocomposites

Graphene (GN) is a two-dimensional monolayer of sp^2-bonded carbon atoms. It has attracted increasing attention in recent years, due to its extraordinarily high electrical and thermal conductivities, great mechanical strength, large specific surface area, and potentially low manufacturing cost. The excellent properties of high specific surface area (2675 m^2 g^{-1}) and high electrical conductivity have made it a suitable material for supercapacitor applications. Use of thermally exfoliated GN nanosheets as supercapacitor electrode materials has been reported to give a maximum specific capacitance of 117 F g^{-1} in aqueous H_2SO_4 electrolyte. For supercapacitors made of chemically modified GN, a specific capacitance of 135 F g^{-1} in aqueous KOH electrolyte has been reported. However, when drying GN during electrode preparation process, the irreversible agglomeration and restacking of GN due to van der Waals interactions to form graphite becomes a major problem for GN based supercapacitors. The agglomeration adversely affects supercapacitor performance by preventing electrolyte from penetrating into the layers. This problem can be avoided by the introduction of spacers into the GN layers. CNTs, metal oxides and conducting polymers can be used as the spacers. Spacers can ensure high electrochemical utilization of GN layers; in addition, electroactive spacers also contribute to the total capacitance. In this section, recent developments on GN-based nanocomposite materials for supercapacitor applications will be reviewed.

Figure 11. SEM micro-images of MnO2-PANI nanocomposites synthesized by (a) interfacial synthesis and (b) chemical co-precipitation (from ref. 30).

3.4.1. Graphene and metal oxide nancomposites

Metal oxides such as CeO_2, RuO_2, V_2O_5 and SnO_2 were used to composite with GN to form advance nanocomposite for supercapacitor applications. Synergistic effect contributed from GN and metal oxide due to improved conductivity of metal oxide and better utilization of GN is expected to contribute to enhance the pseudocapacitance. Jaidev et al. [36] prepared $RuO_2 \bullet xH_2O$-GN nanocomposite by hydrothermal treatment of GN nanosheets, synthesized via exfoliation of graphite oxide in hydrogen environment, with ruthenium chloride in a Teflon-lined autoclave. A symmetrical supercapacitor was fabricated using electrodes prepared by mixing the as-prepared $RuO_2 xH_2O$-GN, activated carbon and Nafion (binder) on conducting carbon fabric. The hybrid nanocomposite shows a maximum specific capacitance of 154 F g^{-1} and energy density of about 11Wh kg^{-1} at a specific discharge current of 1 A g^{-1} (20 wt.% Ru loading). The composite also shows a maximum power density of 5 kW kg^{-1} and coulombic efficiency of 97% for a specific discharge current of 10 A g^{-1}. CeO_2 was also deposited onto the 3D GN by chemical precipitation of 3D GN materials contained $Ce(NO_3)_3$ solution with KOH [37]. CeO_2-GN nanocomposite gave high specific capacitance (208 F g^{-1} or 652 mF cm^{-2}) and long cycle life although the specific surface area of the composite decreases as compared with pure GN. Bonso et al. synthesized GN-V_2O_5 nanocomposite by mixing V_2O_5 sol with the GN/ethanol dispersion and stirred for many days [38]. The thus-prepared GN-V_2O_5 composite electrode achieved specific capacitance value of 226 F g^{-1} in 1 M LiTFSI in acetonitrile. In contrast, the specific capacitance of just V_2O_5 was 70 F g^{-1} and just GN was 42.5 F g^{-1}, demonstrating the synergistic effect of combining the two materials

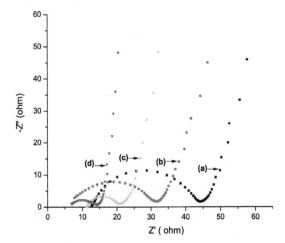

Figure 12. Typical Nyquist impedance plot at open circuit potential (OCP) over a frequency range of 100 kHz–10 mHz with a potential amplitude of 5mV for (a) PEDOT-Aq, (b) PEDOT-Org, (c) nano- $NiFe_2O_4$ and (d) PEDOT– $NiFe_2O_4$ composite electrodes (from ref. 32)

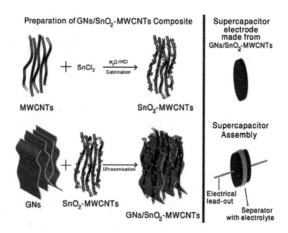

Figure 13. Schematic of preparation of supercapacitor electrode material (from ref. 39).

Transition metal oxide nanoparticles loaded CNTs has been demonstrated as an excellent electroactive material for supercapacitor applications. It is expected that a nanocomposite obtained by dispersion of metal oxide nanoparticles loaded multi-walled CNTs (MWCNTs) into GN can be a good electroactive materials for supercapacitors particularly to increase their cycling stability due to more open structures. Rakhi et al. [39] has prepared the GN-SnO_2/CNTs nanocomposite by ultrasonically mixing of chemically functionalized GN and SnO_2–CNTs. The SnO_2/CNTs was first prepared by chemical precipitation of SnO_2 from $SnCl_2$ solution containing functionalized multi-walled CNTs. The SnO_2/CNTs precipitate was filtrated, washed, dried and calcined. It was then mixed with functionalized GN by ultrasonication to obtain a homogeneous GN-SnO_2/CNTs suspension. Finally, the solid was filtered, washed and dried in a vacuum. To produce chemically functionalized graphene, GN was dispersed in concentrated nitric and sulphuric acid mixture. The process for preparation of GN-SnO_2/CNTs composite is illustrated in Figure 13. The TEM images of multiwall CNTs and SnO_2/CNTs are shown in Figure 14 (a) and (b) respectively. Multi-walled CNTs have an average inner diameter of 10 nm, an outer diameter of 30 nm and an average length in the range of 10–30 μm. Figure 14(b) suggests an uniform distribution of SnO_2 nanoparticles over the surface of multi-walled CNTs. High resolution TEM image of SnO_2/CNTs (inset of Figure 14(b)) reveals that the SnO_2 nanoparticles are highly crystalline in nature with an average particle size of 4–6 nm. TEM images of large area GN and GN-SnO_2/CNTs composite are shown in Figure 14(c) and (d) respectively. SnO_2/CNTs are seen to occupy the surface of GN. Symmetric supercapacitor devices were fabricated by the authors using GN and GN-SnO_2/CNTs composite electrodes. The latter gave remarkable results with a maximum specific capacitance of 224 F g−1, power density of 17.6 kW kg^{-1} and an energy density of 31 Wh kg^{-1}. The results demonstrated that dispersion of metal oxide loaded multi-walled CNTs improved the capacitance properties of GN. The fabricated supercapacitor device exhibited excellent cycle life with ~81% of the initial specific capacitance retained after 6000 cycles.

3.4.2. Graphene and polymer nancomposites

Conducting polymers such as PANI and PEDOT were composited with GN to improve the electrochemical performance of GN for supercapacitor applications. GN-PANI nanocomposite was chemically synthesized by oxidative polymerization of aniline monomer using ammonium peroxydisulfate [$(NH_4)_2S_2O_8$)] as the oxidizing agent in the GN and aniline mixing solution [40]. The presence of GN in polyaniline shows the penetrating network like structure in GN–PANI nanocomposite film, whereas the GN platelets are making the network structure with polyaniline. The high specific capacitance and good cyclic stability have been achieved using 1:2 aniline to GN ratio by weight. The result of Gómeza et al. [40] has proved that the presence of GN in network of polyaniline changes the composite structure. The supercapacitor fabricated using GN–PANI shows the specific capacitance of 300–500 F g^{-1} at a current density of 0.1A g^{-1}.

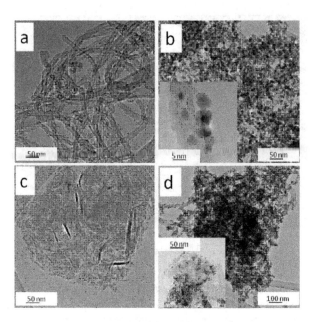

Figure 14. TEM images of (a) MWCNTs, (b) SnO$_2$–MWCNTs (inset shows HRTEM image), (c) GNs and (d) GNs/SnO2–MWCNTs composite (from ref. 39)

GN-PANI composite film with layered structure was obtained via filtration of an aqueous dispersion consisting of positively charged PANI nanofibres and negative charged chemically converted GN sheets that form a stable composite dispersion via electrostatic interaction with the assistance of ultrasonication [41]. The conductivity of GN-PANI film was one order higher than that of pure PANI nanofibres film. The symmetric supercapacitor device using GN-PANI films exhibited a high capacitance of 210 F g^{-1} at 0.3 A g^{-1}, and this capacitance can be maintained for about 94% (197 F g^{-1}) as the discharging current density was increased from 0.3 to 3 A g^{-1}. Due to the synergic effect of both components, the performance of GN-PANI based capacitor is much higher than those of the supercapacitors based on pure chemically converted GN or PANI-nanofibre films. The GN-PANI film has a layered structure as shown in its cross-section scanning electron micrograph (SEM) of Figure 15 (a), which is probably caused by the flow assembly effect of GN sheets during filtration. The magnified SEM image (Figure 15(b)) reveals that PANI nanofibres are sandwiched between chemically converted GN layers. The interspaces between the chemically converted GN layers are in the range of 10-200 nm. This morphology endows GN-PANI film with additional specific surface area comparing with that of the compact GN film prepared under the same conditions (Figure 15c). Filtrating of the dispersion PANI-nanofibres also produced a porous film as shown in Figure 15d, however, the mechanical property of this film is poor and it usually breaks into small pieces after drying.

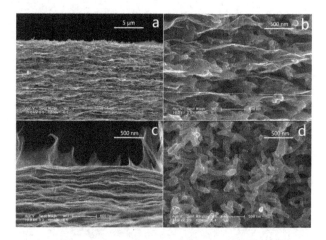

Figure 15. Cross-section SEM images of GN-PANI (a, b), pure chemically converted GN (c), and PANI nanofibre (d) films prepared by vacuum filtration (from ref. 41)

GN was also composited with PEDOT by F. Alvi et al. [42], by chemically oxidative polymerization of ethylene dioxythiophene (EDOT) using ammonium peroxydisulfate [$(NH_4)_2S_2O_8$)] and FeCl$_3$ as oxidizing agents. In the solution of EDOT monomer, GN was added into it at EDOT to GN ratio of 1:1. The GN-PEDOT nanocomposite dramatically improves the electrochemical performance comparing to PEDOT based supercapacitors. The GN-PEDOT has also provided faster electrochemical reaction with an average capacity of 350 F g^{-1}. All the above results been demonstrated that the improvement in supercapacitor performance of GN based electroactive material can be achieved by compositing with metal oxides or conducting polymers. GN in those nanocomposites can act as nanoscale supports for dispersing metal oxides or conducting polymers to increase their surface area. GN can also provide the electronic conductive channels for metal oxides and conducting polymers. In addition, GN nanosheets can restrict the mechanical deformation of the polymers during the redox process due to its unique structural and mechanical properties. Graphene-based nanocomposites are expected to have great future for their application in supercapacitors.

4. Conclusion and future directions

Nanocomposite electroactive materials that have been developed so far have demonstrated huge potential for supercapacitor applications. Different types of nanocomposite electroactive materials, such as mixed metal oxides, polymers mixed with metal oxides, carbon nanotubes mixed with polymers, or metal oxides, and graphene mixed with metal oxides or polymers, can be fabricated by various processes such as solid state reactions, mechanical mixing, chemical co-precipitation, electrochemical anodic deposition, sol-gel, in situ polymerization and other wet-chemical synthesis. It has been shown that significant improvement in term of specific surface area, electrical and ionic conductivities; specific capacitance, cyclic stability, and energy and power density, of supercapacitors can be achieved by using nanocomposite electroactive materials. This can be attributed to the complementary and synergy behaviours of the consisting material components, the unique interface characteristics and the significant increase in surface areas, as well as nano-scale dimensional effects. Electrochemical double layer capacitors using carbon based electroactive materials and pseudocapacitors using metal oxide or conducting polymers as electroactive materials are the two types of most common supercapacitor structures. Compared with the electrochemical double layer capacitors, pseudocapacitors present a number of advantages such as high energy density and low materials cost, but suffer from poor cyclic stability and lower power density. Asymmetric supercapacitors using one electrode of pseudocapacitive materials and the other carbon based double capacitive materials are of great interest and are the current research focus. Nanocomposite pseudocapactive materials have great potential for asymmetric supercapacitor applications. The key issue is to fully utilize nanocomposites' excellent intrinsically properties, especially their high surface area and high conductivity, and to improve the synergistic effect of different electroactive components.

Although nanocomposite films have demonstrated their great potential for supercapacitor applications, several challenges still remain. Synthesis of nanocomposite electroactive materials with precisely controlling their chemical composition ratio, micro/nanostructures, phases, surface area and interfacial characteristics is still challenging. Depending on the preparation technique and process parameters, the property and behaviours of the nanocomposite electroactive materials can vary significantly; therefore, the ability to reproducibly synthesize nanocomposite materials with consistent properties is very important for their wide uses in supercapacitors. Degradation of the nanocomposite electroactive materials stemming from aggregation of the nano-scale components due to the relatively strong forces between them, micro/nanostructure changes due to charging-discharging cycling and materials contaminations due to impurity introduced for original reactants or during synthesis processes etc. has to be resolved before their large-scale adoption by the industry. Most importantly, the costs of materials and their synthesis processes have to be reduced significantly. With the increase in interest and intensive research and development, it is expected that, nanocomposite electroactive materials will have a promising future and will bring a huge change to the energy storage industries.

Acknowledgements

The author would like to thank Transport Canada and National Research Council of Canada (NRC) for supporting the publication of this chapter. The author is also indebted to his NRC colleagues: Dr. Sylvain Pelletier and Ms. Nathalie Legros for their initiation of supercapacitor research at NRC, and Dr. Alexis Laforgue, Dr. Lucie Robitaille, Dr. Yves Grincourt, Dr. Lei Zhang, and Dr. Jiujun Zhang for their collaboration in the supercapacitor research project. He would also like to thanks his research team members: Mr. Brian Gibson, Mr. Marco Zeman, Mr. Robinet Romain, Mr. Benjamin Tailpied, and Ms. Gaëlle LEDUC for their dedication to supercapacitor research. Thank is also given to Ms. Catherine Yang for her editorial review on this chapter.

Author details

Dongfang Yang*

Address all correspondence to: Email: dongfang.yang@nrc.gc.ca

National Research Council Canada, 800 Collip Circle, London, Ontario,, Canada

References

[1] Jiang, H., Li, C., Sun, T., & Ma, J. (2012). High-performance supercapacitor material based on Ni(OH)$_2$ nanowire-MnO$_2$ nanoflakes core–shell nanostructures. *Chem. Commun.*, 48, 2606-2608.

[2] Kim, H., & Popov, N. (2003). Synthesis and Characterization of MnO2Based Mixed Oxides as Supercapacitors. *Journal of the Electrochemical Society*, 150(3), D56-D62.

[3] Lee, M. T., Chang, J. K., Hsieh, Y. T., & Tsai, W. T. (2008). Annealed Mn-Fe binary oxides for supercapacitor applications. *Journal of Power Sources*, 185-1550.

[4] Ye, Z. G., Zhou, X. L., Meng, H. M., Hua, X. Z., Dong, Y. H., & Zou, A. H. (2011). The Electrochemical Characterization of Electrochemically Synthesized MnO2 based Mixed Oxides for Supercapacitor Applications. *Advanced Materials Research*, 287(290), 1290-1298.

[5] Yang. (2012). Pulsed laser deposition of cobalt-doped manganese oxide thin films for supercapacitor applications. *Journal of Power Sources*, 198-416.

[6] Yang, D. (2011). Pulsed laser deposition of manganese oxide thin films for supercapacitor Applications. *Journal of Power Sources*, 196-8843.

[7] Zhong, J. H., Wang, A. L., Li, G. R., Wang, J. W., Ou, Y. N., & Tong, Y. X. (2012). Co_3O_4/$Ni(OH)_2$ composite mesoporous nanosheet networks as a promising electrode for supercapacitor applications. *J. Mater. Chem.*, 22-5656.

[8] Yang, Y., Kim, D., Yang, M., & Schmuki, P. Vertically aligned mixed 2O5-TiO2 nanotube arrays for supercapacitor applications. 2011, *Chem. Commun.*, 47, 7746-7748.

[9] Jayalakshmi, M., Venugopal, N., Phani, Raja. K., & Mohan, Rao. (2006). Nano SnO2 - Al_2O_3 mixed oxide and SnO_2-Al_2O_3-carbon composite oxides as new and novel electrodes for supercapacitor applications. *Journal of Power Sources*, 158, 1538-1543.

[10] Wang, C., Zhang, X., Zhang, D., & Yao, C. (2012). Facile and low-cost fabrication of nanostructured NiCo2O4 spinel with high specific capacitance and excellent cycle stability. *Electrochimica Acta*, 63-220.

[11] Chang, S. K., Lee, K. T., Zainal, Z., Tan, K. B., Yusof, N. A., Yusoff, W., Lee, J. F., & Wu, N. L. (2012). Structural and electrochemical properties of manganese substituted nickel cobaltite for supercapacitor application. *Electrochimica Acta*, 67.

[12] An, K. H., Jeon, K. K., Heo, J. K., Lim, S. C., Bae, D. J., & Lee, Y. (2002). High-Capacitance Supercapacitor Using a Nanocomposite Electrode of Single-Walled Carbon Nanotube and Polypyrrole. *Journal of The Electrochemical Society*, 149(8), A1058-A1062.

[13] Deng, M., Yang, B., & Hu, Y. (2005). Polyaniline deposition to enhance the specific capacitance of carbon nanotubes for supercapacitors. *Journal of Materials Science Letters*.

[14] Tadai, K., & Mitani, T. (2005). Highly dispersed ruthenium oxide nanoparticles on carboxylated carbon nanotubes for supercapacitor electrode materials. *J. Mater. Chem.*, 15-4914.

[15] Kim, Y., & Mitani, T. (2006). Oxidation treatment of carbon nanotubes: An essential process in nanocomposite with RuO_2 for supercapacitor electrode materials. *Appl. Phys. Lett.*, 89, 033107.

[16] Hsieh, T. F., Chuang, C. C., Chen, W. J., Huang, J. H., Chen, W. T., & Shu, C. M. (2012). Hydrous ruthenium dioxide/multi-walled carbon-nanotube/titanium electrodes for supercapacitors. Oxidation treatment of carbon nanotubes:. *An essential process in nanocomposite with RuO2 for supercapacitor electrode materials*, 50, 1740-1747.

[17] Lee, J. K., Pathan, H. M., Jung, K. D., & Joo, O. S. (2006). Electrochemical capacitance of nanocomposite films formed by loading carbon nanotubes with ruthenium oxide. *Journal of Power Sources*, 159-1527.

[18] Kim, B. C., Wallace, G. G., Yoon, Y. I., Ko, J. M., & Too, C. (2009). Capacitive properties of RuO_2 and Ru-Co mixed oxide deposited on single-walled carbon nanotubes for high-performance supercapacitors. *Synthetic Metals*, 159-1389.

[19] Chen, Y. M., Cai, J. H., Huang, Y. S., Lee, K. Y., & Tsai, D. (2011). Preparation and characterization of iridium dioxide-carbon nanotube nanocomposites for supercapacitors. *Nanotechnology*, 22(115706), 7.

[20] Lee, C. Y., Tsai, H. M., Chuang, H. J., Li, S., Lin Yi, P., & Tseng, T. (2005). Characteristics and Electrochemical Performance of Supercapacitors with Manganese Oxide-Carbon Nanotube Nanocomposite Electrodes. *Journal of The Electrochemical Society*, 152(4), 716A-A720.

[21] Lee, J. Y., Liang, K., An, K. H., & Lee, Y. (2005). Nickel oxide/carbon nanotubes nanocomposite for electrochemical capacitance. *Synthetic Metals*, 150-153.

[22] Liu, M. Y., Hsieh, T. F., Hsieh, C. J., Chuang, C. C., & Shu, C. M. (2011). Performance of vanadium oxide on multi-walled carbon nanotubes/titanium electrode for supercapacitor application. *Advanced Materials Research*, 311-313, 414-418.

[23] Jayalakshmi, M., Mohan, Rao. M., Venugopal, N., & Kim, K. B. (2007). Hydrothermal synthesis of SnO_2-2O5 mixed oxide and electrochemical screening of carbon nanotubes (CNT), V_2O_5, V_2O_5-CNT, and SnO_2-V_2O_5-CNT electrodes for supercapacitor applications. *Journal of Power Sources*, 166, 578-583.

[24] Lee, B. J., Sivakkumar, S. R., Ko, J. M., Kim, J. H., Jo, S. M., & Kim, D. (2007). Carbon nanofibre/hydrous RuO_2 nanocomposite electrodes for supercapacitors. *Journal of Power Sources*, 168-546.

[25] Cormier, Z. R., Andreas, H. A., & Zhang, P. (115). Temperature-Dependent Structure and Electrochemical Behaviour of RuO_2/Carbon Nanocomposites. *J. Phys. Chem*, 115, 19117-19128.

[26] Selvakumar, M., Bhat, D. K., Aggarwal, A. M., Iyer, S. P., & Sravani, B. (2010). *Physica*, 405-2286.

[27] Tao, T., Chen, Q., Hu, H., & Chen, Y. (2012). MoO_3 nanoparticles distributed uniformly in carbon matrix for supercapacitor applications. *Materials Letters*, 66-102.

[28] Chen, L., Sun, L., Luan, F., Liang, Y., Li, Y., Liu, X., & , X. (2010). Synthesis and pseudocapacitive studies of composite films of polyaniline and manganese oxide nanoparticles. *Journal of Power Sources*, 195-3742.

[29] Hu, Z., Xie, Y., Wang, Y., Mo, L., Yang, Y., & Zhang, Z. (2010). Synthesis and pseudocapacitive studies of composite films of polyaniline and manganese oxide nanoparticles. *Journal of Power Sources*, 195-3742.

[30] Wang, J., Yang, Y., & Huang, Z. (2012). . Interfacial synthesis of mesoporous MnO2/polyaniline hollow spheres and their application in electrochemical capacitors. *Journal of Power Sources*, 204-236.

[31] Saranya, S., & Selvan, R. K. (2012). Synthesis and characterization of polyaniline/$MnWO_4$ nanocomposites as electrodes for pseudocapacitors. *Applied Surface Science*, 258-4881.

[32] Sen, P. (2010). Electrochemical performances of poly 3 4ethylenedioxythiophene)-$NiFe_2O_4$ nanocomposite as electrode for supercapacitor. *Electrochimica Acta*, 55, 4677-4684.

[33] Murugan, A. V., Viswanath, A. K., & Gopinath, C. S. (2006). Highly efficient organic-inorganic poly (3 4-ethylenedioxythiophene)-molybdenum trioxide nanocomposite electrodes for electrochemical supercapacitor. *J. Appl. Phys.*, 100(074319), 1-5.

[34] Dong, Z. H., Wei, Y. L., Shi, W., & Zhang, G. (2011). Characterisation of doped poly-pyrrole/manganese oxide nanocomposite for supercapacitor electrodes. *Materials Chemistry and Physics*, 131-529.

[35] Yang, X., Wang, G., Wang, R., & Li, X. (2010). A novel layered manganese oxide/poly (aniline-co-o-anisidine) nanocomposite and its application for electrochemical super-capacitor. *Electrochimica Acta*, 55-5414.

[36] Jaidev, Jafri. R. I. (2011). Ramaprabhu S. Hydrothermal synthesis of $RuO_2.xH_2O$/graphene hybrid nanocomposite for supercapacitor application. International Con-ference on Nanoscience,. *Technology and Societal Implications*, 8-10.

[37] Wang, Y., Guo, C. X., Liu, J., Chen, T., Yang, H., & Li, C. (2011). CeO_2 nanoparticles/graphene nanocomposite-based high performance supercapacitor. *Dalton Trans.*, 40(6388).

[38] Bonso, J. S., Rahy, A., Perera, S. D., Nour, N., Seitz, O., Chabal, Y. J., Balkus Jr, K. J., Ferraris, J. P., & Yang, D. (2012). Exfoliated graphite nanoplatelets-2O5 nanotube composite electrodes for supercapacitors. *Journal of Power Sources*, 203, 227-232.

[39] Rakhi, R. B., & Alshareef, H. N. (2011). Enhancement of the energy storage properties of supercapacitors using graphene nanosheets dispersed with metal oxide-loaded carbon nanotubes. *Journal of Power Sources*, 196-8858.

[40] Gómeza, H., Ramb, M. K., Alvia, F., Villalba, P., & Stefanakos, E. (2011). Graphene-conducting polymer nanocomposite as novel electrode for Supercapacitors. *Journal of Power Sources*, 196-4102.

[41] Wu, Q., Xu, Y., Yao, Z., & Liu, A. (2010). Supercapacitors Based on Flexible Gra-phene/Polyaniline Nanofiber Composite Films. ACS. *Nano*, 4(4), 1963-1970.

[42] Stefanakos, E., Goswami, Y., & Kumar, A. (2011). Graphene-polyethylenedioxythio-phene conducting polymer nanocomposite based supercapacitor. *Electrochimica Acta*, 56-9406.

Chapter 5

Photonics of Heterogeneous Dielectric Nanostructures

Vladimir Dzyuba, Yurii Kulchin and
Valentin Milichko

Additional information is available at the end of the chapter

1. Introduction

Over last 20 years great scientific attention has been paid to nanostructures and nanocomposites based on nanoparticles of semiconductor materials ($1eV<E_{gap}<3eV$) since they exhibit a wide range of nonlinear properties and can be used in various applied fields [1-5]. However, the history of active investigation of dielectric nanostructures' properties started only recently [6-8]. These structures are the heterogeneous medium formed by liquid or solid dielectric matrices (e.g., polymer glasses and oils) and nanoparticles of dielectrics (Al_2O_3, SiO_2, MgO, etc.).

As some experiments have shown [6,8,9], such structures have nonlinear optical properties, whose dependence on the intensity and optical radiation wavelength is not typical of previously known nonlinear optical media [10-14]. The anomalous nonlinear optical properties are manifested in the fact that, firstly, the nanostructures' optical response on the radiation is of a non-thermal nature and occurs at radiation intensities below $1kW/cm^2$ [9]. Second, despite the wide band gap of nanostructures' components ($E_{gap}>3eV$), this response takes place in the visible and infrared region of light spectrum, and reaches a maximum then decreases to zero under increasing intensity [8,9,15-17].

The dielectric nanostructures show other unexpected nonlinear optical properties. The nonlinear interaction of high-intensity radiation of different frequencies results in the generation of harmonics in conventional dielectric media. In the case of propagation of the low-intensity radiation of different frequencies in the dielectric nanostructures, the nonlinear interaction is manifested in the dependence of the light beam intensity on the intensity of another collinearly propagating light beam [18]. The two-frequency interaction observed in nanostructures does not prevent the generation of harmonics, but this process requires radiation intensities four orders of magnitude higher.

The optical nonlinearity of dielectric nanostructures allows one to believe that they will be used to develop and create new optoelectronic [19-24] and fibre-optic devices to control [11], process and transmit the information [25]. Of no less interest are the prospects of using such nanostructures in new optical materials with controlled optical properties, in particular, photonic crystals [26] and the media generating optical solitons at low intensities. In addition, several international research groups have proposed using these structures to create the elements of electrical circuits, since the nonlinear properties appear in the range of THz and GHz radiation and under the influence of an electric potential [27-30].

The study of nonlinear optical properties of dielectric nanostructures containing nanoscale objects of different chemical natures, shapes and sizeshas shown that the existence of a low-threshold optical response is due to a number of conditions. The first is the presence of defect levels in the band gap of nanoparticles' charge carriers and this is manifested in the form of absorption bands in the nanoparticles' transmission spectrum [8,9,31-33]. Second, the radiation forming the nonlinear response of the nanostructure must have a frequency lying within the absorption band [8,9]. Third, the size and shape of the nanoparticles have to lead to the formation of a wide range of exciton states due to the quantum size effect [10,34-38]. Fourth, the matrix permittivity must be less than that of nanoparticle material, since the chemical nature of the matrix material significantly affects the formation of long-lived exciton states [9,31,39-42]. Fifth, the value of electric dipole moments induced by electrons phototransition should be substantially larger than dipole moments in the bulk material. It allows observing the optical nonlinearity of nanostructures with a low concentration of nanoparticles under low-intensity optical fields.

The theoretical description of the observed effects [43-45] is based on the fact that the occurrence of nontypical optical nonlinearity requires the existence of defect levels and the broad band of exciton states in the energy band gap of charge carriers. The radiation causes the electron transitions from the defect to the exciton levels, thereby creating the photo-induced population difference. This process is accompanied by the appearance of the nanoparticle electric dipole moment, herewith its module depends nonlinearly on the intensity and light wavelength. The theory conclusions and theoretical modelling of transmission spectrum and the behaviour of the nonlinear refractive index are very similar to the experimental results [9].

It follows from the theory that the nature of the nonlinearity is determined not only by the behaviour of the photo-induced dipole moment module in an external field, but also by the nanoparticle's orientation along the vector E. However, this orientation has a minor contribution to the nonlinearity, so the observed nonlinear optical response can take place in the case of unpolarized light and solid nanostructures, which is in agreement with the experiment.

This chapter is an original quantitative study of the nonlinear refraction and absorption of continuous low-intensity laser radiation in different heterogeneous dielectric nanostructures and compares these data with theoretical ones. In addition, the theory of nonlinear light transmission by dielectric nanostructures is discussed.

2. Experimental Chapter

2.1. Dielectric Nanocomposite Preparation and Spectral Features

To study the changes of optical characteristics of the heterogeneous dielectric nanostructuresunder continuous low-intensity radiation we used the dielectric Al_2O_3, SiO_2, TiO_2 and ZnO nanoparticles with 7,2; 8; 3,4 and 3,3eV band gaps of the bulk samples respectively [46]. The nanoparticles were purchased from Sigma Aldrich Company and investigated by AFM microscopy (Figures 2,3). The Al_2O_3 nanoparticles were used from work [9]. The averaged dimensions of nanoparticles are 45nm in diameter and 6nm in height for Al_2O_3, 20nm in diameter and 10nm in height for SiO_2, 15nm in diameter and 5nm in height for TiO_2 and 50nm in diameter and 40nm in height for ZnO. As a matrix for nanoparticles, dielectric immersion and transformer oils were used. The immersion oil consists of weak polar molecules and is based on cedar resin witha negative temperature gradient of refractive index $|dn/dT|=4*10^{-4}$. The transformer oil is a PDMS (polydimethylsiloxane) liquid having an optical transparency up to 200nm, chemical inertness, a high heat resistance ($|dn/dT|<10^{-7}$) and high stability of dielectric characteristics.

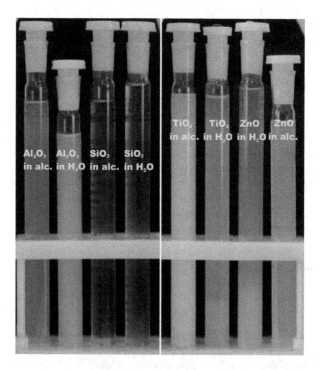

Figure 1. The nanoparticles suspension in isopropyl alcohol and distilled water.

The nanopowders were dissolved in isopropyl alcohol to precipitate the particles sticking together (Figure 1). After precipitation the upper isopropyl alcohol layers containing separate nanoparticles with small deviation in size were added into the oil (Al_2O_3 into immersion oil; SiO_2, TiO_2 and ZnO into PDMS) pre-heated to 40°C. Slow heating of the mixture to 60°C resulted in the appearance of convection currents which, in turn, form the uniform nanoparticles' distribution over the entire suspension volume and lead to alcohol evaporation. Then heterogeneous dielectric nanostructures (hence forth called the HDN) with nanoparticle volume concentration >1% were placed in a quartz cuvette 5 mm thick and 18.7 mm in length.

The nonlinear response of the medium on the radiation of certain frequencies can take place if this medium has nonlinear spectral characteristics that are related directly to the energy spectrum structure of charge carriers. When dealing with a nanoparticle we can expect the energy spectrum of its charge carriers to depend on the form and degree of nanoparticle surface development. Besides, the energy spectrum will depend on the matrix material and, to a greater extent, on its permittivity ε. Therefore, we studied the transmission spectra of Al_2O_3 (permittivity of the bulk sample is $\varepsilon_{stat}=10$), SiO_2 ($\varepsilon_{stat}=4,5$), TiO_2 ($\varepsilon_{stat}>86$) and ZnO ($\varepsilon_{stat}=8,8$) nanoparticles suspended in isopropyl alcohol ($\varepsilon_{stat}=24$), distilled water ($\varepsilon_{stat}=80$) and oil ($\varepsilon_{stat}=2,5$).

As it follows from the transmission spectra figures (Figure 4), the nanoparticles of broadband dielectrics (Al_2O_3 and SiO_2) suspended in oil have a non-symmetric broad absorption band that is formed by exciton states with high density. The asymmetry of the absorption band is explained due to the broadening of the exciton levels. This band is not observable either in the bulk sample or in the nanoparticles' array suspended in other media.This can be explained by the fact that the electronic structure of nanoparticles embedded in a matrix depends strongly on the ratio between the permittivity of matrix ε_1 and nanoparticles ε_2 [39,40]. Given $\varepsilon_1/\varepsilon_2>1$, polarization interaction leads to attraction of positive charges to the inner surface of the nanoparticles and to the destruction of defect states by virtue of interaction the nanoparticle electrons from these levels with high-polarized matrix molecules. If $\varepsilon_1/\varepsilon_2<1$, polarization interaction causes repulsion of charges from the nanoparticle's surface into the interior, thus preserving these states. So, the propagation of visible radiation ($2,1eV<E<3,1eV$) into the HDN based on oil and Al_2O_3 nanoparticles results in electron transition from defect to exciton levels (Figure 5). In the case of water and alcohol matrixes there are no free electrons on the defect level, therefore, electron transition under ultraviolet radiation ($E>4eV$) is available. The same situation is exists for the HDN based on oil and SiO_2 nanoparticles: the exciton generation occurs under visible and ultraviolet radiation ($2eV<E<6eV$), however, less probable electron transitions from valence band to exciton levels can take place provided $E>2eV$ (Figure 5).

The TiO_2 and ZnO nanoparticles' array transmission spectra (Figure 6) inform that the nanoparticles of narrow-band dielectrics suspended in oil have a blurred edge of fundamental absorption that is formed by exciton states without any absorption band within (400-700)nm. The matrix permittivity effects only the position of fundamental absorption.

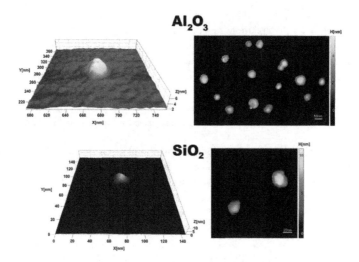

Figure 2. AFM images of Al$_2$O$_3$ and SiO$_2$ nanoparticles precipitated on a mica place. Defined dimensions are 45nm in diameter and 6nm in height for Al$_2$O$_3$, 20nm in diameter and 10nm in height for SiO$_2$.

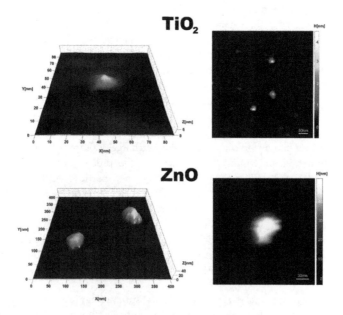

Figure 3. AFM images of TiO$_2$and ZnOnanoparticles precipitated on a mica place. Defined dimensions are15nm in diameter and 5nm in height for TiO$_2$, 50nm in diameter and 40nm in height for ZnO.

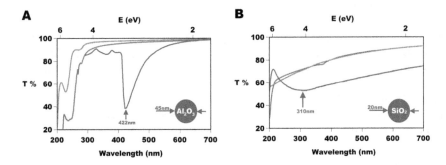

Figure 4. Transmission spectra of Al_2O_3 (A) and SiO_2 (B) nanoparticles' array dispersed in H_2O (green curve), isopropyl alcohol (red curve) and oil (blue curve). The curves were obtained by division of the T spectrum of nanoparticle suspension by that of the matrix.

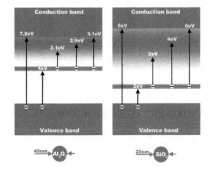

Figure 5. The energy band gap structure of Al_2O_3 and SiO_2 nanoparticles' array.

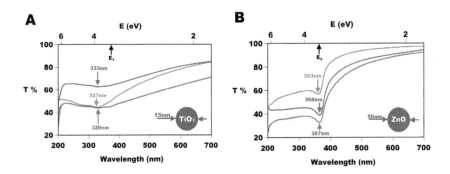

Figure 6. Transmission spectra of TiO_2 (A) and ZnO (B) nanoparticles' array dispersed in H_2O (green curve), isopropyl alcohol (red curve) and oil (blue curve). The curves were obtained by division of the T spectrum of nanoparticle suspension by that of the matrix.

2.2. Z-scan Experiment

The experimental study of the dependence of the HDN nonlinear optical response on the laser radiation of variable intensity was performed using the standard z-scan technique with open and closed apertures [47,48] (Figure 7). Since the nonlinear optical response of the HDN appears under low-intensity optical fields we used the semiconductor lasers providing focal intensities of up to 500 W/cm^2 for green and violet radiation. Such low intensities are at least four orders of magnitude lower than the pulsed mode intensity required for appearance of the nonlinear response in previously known environments [10-14].

In the z-scan experiments, the HDN based on Al$_2$O$_3$, SiO$_2$, TiO$_2$ and ZnO nanoparticles can be considered as thin experimental samples, since the interaction length of radiation with the samples is equal to L=5mm, being less than minimum Rayleigh range [47] $Z_0=(\pi\omega_0^2)/\lambda$=8.7mm.

The values the changes in refractive index Δn (I,λ) and absorption coefficient $\Delta\alpha(I, \lambda)$ one can calculate by the following:

$$\Delta n(I, \lambda) = \frac{\lambda \Delta T_{pv} \ln^{I_0}/_I}{0.812\pi(1-S)^{0.27}L\left(1-^I/_{I_0}\right)}$$ (1)

$$\Delta a(I, \lambda) = \frac{2\sqrt{2}\Delta T}{L}$$ (2)

that has been derived from the number of expressions:

$$\Delta T_{pv} \approx 0.406(1-S)^{0.27}\,|\,\Delta\Phi_0\,|$$ (3)

$$\Delta\Phi_0 = \frac{2\pi}{\lambda}n_2I_0L_{eff}$$ (4)

where λ is a radiation wavelength and I_0 and I are the input and output intensities, respectively. S is a fraction radiation transmitted by the aperture in the absence of the sample (S=0.04 and 0.06 provided the HDN based on Al$_2$O$_3$ in green and violet optical field, respectively; S=0.22 and 0.35 provided the HDN based on SiO$_2$, TiO$_2$, ZnO in green and violet optical field, respectively), L_{eff}=L*(1-e$^{-\alpha L}$), L and α are the sample length and absorption coefficient, respectively and ΔT is a normalized change in integral transmitted intensity. The ratio of I/I$_0$ was determined due to the transmittance characteristics of the HDN (Figure 8). Since I/I$_0$=(P$_{out}$S$_0$)/(P$_{in}$S), where P$_{in}$ and P$_{out}$ are input and output radiation powers, S$_0$ and S are the beam squares into the sample, and near the outer surface, the ratio P$_{out}$/P$_{in}$ was defined from Figure 8 and S/S$_0$ negligibly exceeds the unite.

Linear behaviour of the HDN transmittance under increasing optical field is not reflected in the real behaviour that was most detailed in the study by z-scan with open aperture and will be discussed below. However, the z-scans have shown the changes of absorption $\Delta T < 5\%$ that is absolutely imperceptible in Figure 8.

It is necessary to clarify that the z-scan of low-intensity continuous radiation gives information about the total impact of all physical processes, excited by radiation in the matrix and nanoparticles, on the optical properties of the HDN.Thus, we have to divide the matrix effect from the nonlinear response caused by presence of nanoparticles.

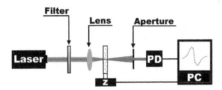

Figure 7. The experimental setup used for z-scan experiments. Setup includes: semiconductor sources of coherent continuous radiation (wavelengths of 532 and 442 nm with maximum power 22 and 35mW and beam diameters 1,1 and 0,95 mm, respectively); the PD-power photodetector; 75mm (focus diameters $\omega_0 = 71$mkm for the green radiation and 92mkm for the violet) and 50mm ($\omega_0 = 46$ mkm for the green radiation and 90 mkm for the violet) lens; Z is z shift and PC is a computer.

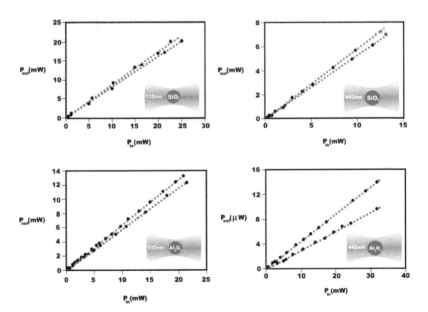

Figure 8. The integral output power P_{out} as a function of input power P_{in} of green and violet radiation propagating in the oil (red curve) and the HDN (blue curve).

The results of measurements of the change in Δn of the HDN based on Al_2O_3 nanoparticles obtained by z-scan are shown in Figure 9. Considering the absorption of radiation, we used z-scans with open aperture [9].The obtained results demonstrate a linear absorption of green radiation in the samples of pure oil and the HDN (normalized transmittance is equal to 1 for all z) within intensity range (0; 300) W/cm².However, the experiments revealed an absorption saturation of violet radiation. These experimental data have been used to adjust the normalized transmittance curves obtained with the use of the closed aperture technique, in accordance with the well-known method [47,48].

The results of the z-scan with closed aperture revealed the negative thermal change in the matrix refractive index and nonlinear negative change in the HDN refractive index. Since all physical processes in the matrix and nanoparticles, forming the change of the HDN refractive index under optical field, can be considered to a first approximation as independent processes $\Delta n_{HDN} = \Delta n_{matrix} + \Delta n_{nanopart}$, so we can assume that $\Delta n (I, \lambda)$ for the HDN is a nonlinear function of input radiation intensity (Figure 9).

To study the effect of the matrix static permittivity on the electronic structure and optical properties of nanoparticles' array, we also investigated Al_2O_3nanoparticles suspended in isopropyl alcohol and distilled water according to the z-scan technique.The experiment showed a total absence of the nonlinear response in these media, being in agreement with our description in the introduction.

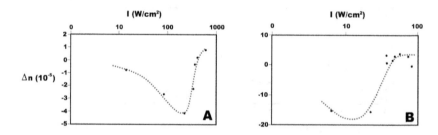

Figure 9. Change of refractive index Δn (I,λ) of the HDN based on Al_2O_3 nanoparticles under green (A) and violet (B) radiation. The curves were obtained by division of Δn for the HDN by Δn for the immersion oil.

The results of measurements of the refractive index and absorption coefficient of the HDN based on SiO_2 obtained by z-scan are shown in Figures 10 and 11. Since the transformer oil characteristics are not significantly changed within intensity range (0;500) W/cm², so the z-scan with close and open aperture informs only about changes in the HDN optical parameters caused by nanoparticles. The obtained results demonstrate the nonlinear refraction of laser radiation (Figure 10A,C and 11A,C) in the HDN within intensity range (0; 300)W/cm². In addition, the experiments revealed the nonlinear absorption of green and violet radiation in the HDN (Figure 10B,D and 11B,D) within the same intensity range. However, the HDN based on the nanoparticles of the narrow-band dielectrics (TiO_2 and ZnO) did not reveal any nonlinear changes in refraction and absorption under continuous low-intensity radiation.

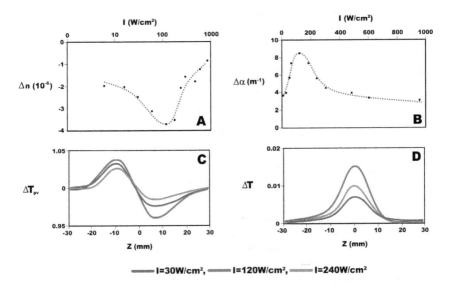

Figure 10. Change of refractive index Δn (A) and absorption coefficient $\Delta\alpha$ (B) of the HDN based on SiO_2 nanoparticles under green radiation; C and D are the approximation of z-scan results with close and open aperture, respectively.

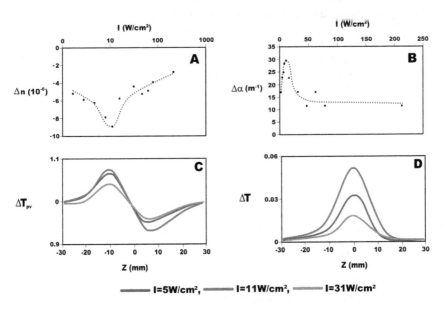

Figure 11. Change of refractive index Δn (A) and absorption coefficient $\Delta\alpha$ (B) of the HDN based on SiO_2 nanoparticles under violet radiation; C and D are the approximation of z-scan results with close and open aperture, respectively.

3. The Theory of The HDN Nonlinear Transmittance

3.1. Preface

The theoretical description of the physical and optical properties of heterogeneous nano-composites containing nanoparticles is a complex problem. In fact, it seems impossible to correctly calculate the physical characteristics of an individual nanoparticle as a system consisting of a great number of particles obeying the quantum mechanics laws. Attempts to apply the well-known methods of solid state physics to describe the nanoparticles' properties run into problems, since it is not possible to disregard the effects caused by surface defects, as well as crystal lattice defects. It is known that the optical properties of a quantum mechanical system are associated with the features of the energy spectrum of charge carriers (electrons and holes).

At the present time, it is beyond all question that the optical and electric properties of nano-particles have wide differences with that of the bulk samples due to the features of the energy spectra. These differences are caused by three effects. First, the band gap of nanoparticle charge carriers contain the allowed energies zone, herewith, the energy structure defined by the high density of surface structural defects and the irregular shape of nanoparticles. Second, the excitons and discrete energy spectra are formed below and into the conduction band due to the small nanoparticle size and size-quantization effect, respectively. In turn, the size quantization effect is caused by spatial confinement of the charge carriers' wave functions. Third, the electric dipole moments of electronic transitions in such quasi zero dimensional systems can be larger than that of the bulk sample. The formation of the above mentioned states is of threshold character, herewith, the threshold depends on the nanoparticle dimensions. Specifically, for a spherical nanoparticle (with the permittivity ε_2) dispersed in a medium (ε_1), such states can be formed if the nanoparticle radius a is smaller than some critical radius a_c:

$$a \leq a_c = 6 \mid \beta \mid^{-1} a_{e,h}$$

(5)

where

$$\frac{\varepsilon_1 - \varepsilon_2}{\varepsilon_1 + \varepsilon_2}$$

(6)

Here $a_{e,h}$ is the Bohr radius of charge carriers in the nanoparticle material [49].

Some properties of the quantum states' spectrum can be clarified by studying the nanocomposites' transmittance spectra. As a rule, experimental studies are concerned with the transmittance spectra of nanoparticles' arrays embedded in a solid matrix or deposited on the transparent material surface. In this case, the electronic structure of nanoparticles is substantially influenced by the matrix material and the interaction between nanoparticles. Because of these effects, it is not possible to consider the transmittance spectra as the spectra of no interacting nanoparticles' arrays. Nanocomposites containing low concentrations of nano-particles almost satisfy the condition for the lack of the above mentioned interactions, how-ever, to study the optical properties of such composites, one cannot take into account the effects of the optical field on the distribution of the particles throughout degrees of freedom. In this case, given the low-intensity radiation, the optical field effect on the coordinates of a gravity centre of a nanoparticle can be disregarded, that cannot be said about the distribu-tion of particles throughout the rotational degrees of freedom.

At the present time, there is no well-known theoretical approach taking into account not on-ly the characteristic of nanoparticle dimensions, but also the orientation of nanoparticles in the external field of laser radiation, the dependences of the scattering and absorption cross sections on the propagating radiation intensity. In this context, it is necessary to develop a theoretical model of the scattering and absorption cross sections in dielectrics nanocompo-sites with the above mentioned features of such systems.

In this study, we suggest a semiphenomenological model of the optical transmittance of the array of noninteracting small sized ($a<a_c$) dielectric nanoparticles embedded in the dielectric matrix. We show that the basic mechanisms of low-threshold effects of nonlinear scattering and absorption of laser radiation in the HDN are: the photo induction of electric dipole mo-ments of nanoparticles in the external optical field; the orientation of nanoparticles along the polarization direction of this field. In addition, we will discuss the behaviour of the HDN transmittance in the central frequency vicinity of the absorption band and the dependence of the band depth on the radiation intensity.

3.2. The Theoretical Approach

We consider the HDN consisting of the low concentration (the number of nanoparticles N per unit volume) of dielectric nanoparticles embedded in an isotropic transparent dielectric matrix with a small coefficient of viscosity and linear optical properties within the visible spectral range. In our case, the multiple scattering of radiation by nanoparticles and the nanoparticles' interaction with each other can be neglected. Let us introduce two coordinate systems with the same origin (Figure 12).

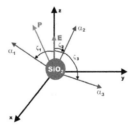

Figure 12. The coordinate system used in the theoretical study.

One of the systems $\{\alpha_1, \alpha_2, \alpha_3\}$ corresponds to the coordinates coinciding with the principle axes of the particle polarization tensor with the unit vectors (n_1, n_2, n_3). The other system is the Cartesian laboratory coordinate system $\{x, y, z\}$ with the unit vectors (n_x, n_y, n_z). We suggest that the electromagnetic wave polarized along the zaxis $E = \{0,0,E\}$ is incident on the composite. We chose the xaxis to be directed collinearly with a wave vector.

The optical transmittance of the HDN depends on the extinction coefficient, the path of the light beam in the material and the optical reflectance from the HDN boundary. For normal incidence of the light beam onto the boundary of the planar nanocomposite layer arranged normally to the xaxis, the transmittance expression can be written as [50]

$$T(\omega, N) = \frac{(1 - R^2)^2 \ e^{\beta L}}{1 - R^2 \ e^{-2\beta L}} \tag{7}$$

Here, β is the extinction coefficient, R is the optical reflectance of the boundary (in experiments R is much smaller than unit) and L is the interaction length of light beam with the HDN.

In the case of single scattering approximation, the extinction coefficient can be expressed in terms of the scattering $\sigma^s(\omega,a)$ and absorption $\sigma^a(\omega,a)$ cross sections of the HDN unit volume as

$$\beta(\omega, a) = \sigma^a(\omega, a) + \sigma^s(\omega, a) + a^m(\omega) \tag{8}$$

where $\alpha^m(\omega)$ is the extinction coefficient of the matrix material and a is the characteristic of nanoparticle dimension. For the above indicated orientation of the nanocomposite layer, the scattering and absorption cross sections in the laboratory coordinate system can be expressed in terms of the polarizability component of the HDN unit volume $\chi_{zz}(\omega,a)$ by the relations [50,51]

$$\sigma^a(\omega, a) = \frac{4\pi\omega}{c} \operatorname{Im}\chi_{zz}(\omega, a)$$
$$d\sigma^s(\omega, a) = \frac{\omega^4}{c^4} |\chi_{zz}(\omega, a)|^2 \sin^2\theta \ d\Omega \tag{9}$$

Here θ is the angle of the vector directed along the scattering direction and cis the light speed in vacuum.

We introduce the effective polarizability tensor for nanoparticle in the matrix $\alpha = \{\alpha_{ij}\}$ in such a way that the components of the nanoparticle electric dipole moment Pinduced by the external plane polarized monochromatic electromagnetic field E with the frequency ω are determined directly in terms of the external field rather than the local field $P_i = \alpha_{ij} E_j$. In the coordinate system $\{\alpha_1, \alpha_2, \alpha_3\}$, the polarization vector of the nanoparticle is

$$P = \sum_j^3 a_{ij} (n_j E) n_j \tag{10}$$

If the vector E is directed along the z axis, the zcomponent of the polarization vector is

$$P_z = \sum_j^3 a_{ij} E (n_j n_z)^2 \sum_j^3 a_{ij} E \cos^2 \theta_j \tag{11}$$

Here θ_j is the angle between vector E and the α_j axis; this angle specifies the nanoparticle orientation in the external electromagnetic field in the laboratory coordinate system. Since the nanoparticles are randomly oriented, we assume that the polarizability tensor of the medium $\chi = \{\chi_{ij}\}$ is diagonal and the polarization vector of the HDN unit volume in the laboratory coordinate system is $P_z = \chi_{zz} E$. Comparing this expression with (11), we obtain

$$\chi_{zz} = N \left(a_{11} \cos^2 \theta_1 + a_{22} \cos^2 \theta_2 + a_{33} \cos^2 \theta_3 \right) \tag{12}$$

After simple transformation, taking into account that $\cos^2 \theta_1 + \cos^2 \theta_2 + \cos^2 \theta_3 = 1$, we can obtain an expression that relates the component of χ_{zz} in the laboratory coordinate system with the diagonal components of the nanoparticle polarizability tensor in the principle axes system:

$$\chi_{zz} = N \left(a_0 + \Delta a_1 Q_1 + \Delta a_2 Q_2 \right) \tag{13}$$

where

$$a_0 = \frac{a_{11} + a_{22} + a_{33}}{3}$$
$$\Delta a_1 = a_{11} - a_{33} \tag{14}$$
$$\Delta a_2 = a_{22} - a_{33}$$

The values averaged over all possible orientations

$$Q_1 = \left\langle \cos^2 \theta_1 - \frac{1}{3} \right\rangle$$
$$Q_2 = \left\langle \cos^2 \theta_2 - \frac{1}{3} \right\rangle \tag{15}$$

are the orientation order parameters of the nanoparticles ensemble in the external field. The angle distribution function of nanoparticles and, hence, the order parameters Q_1 and Q_2, depend on the laser radiation intensity and, via the components α_{ij}, on the radiation frequency. The quantities Q_1 and Q_2 as functions of the intensity exhibit the saturation at I>Ip irrespective of the matrix material.

The low-threshold nonlinear optical response takes place if the transmittance spectrum of the nanoparticles' array exhibits the broad absorption bands lacking in the bulk sample spectrum [31-33]. The polarizability tensor components α_{ij} of the nanoparticle are to reach their maxima corresponding to dipole transitions of charge carriers from the state <n| to the state |g> within this frequency region. In addition, it is known that the diagonal tensor components in the coordinate system of the principal axes within this frequency region can be expressed as [52]

$$a_{jj}(\omega) = \sum_{n,g} \frac{|\langle n \mid er_j \mid g \rangle|^2}{\Box(\omega - \omega_{ng} + i\Gamma_{ng})} \Delta \rho_{ng} \tag{16}$$

The summation is performed over all allowed optical transitions of charge carriers of the nanoparticles with the frequency transition ω_{ng} from the states <n| to the states |g>, being the component of the electric dipole moment of the transition $p^j{}_{ng}$=<n|er_j|g> and the transitions width Γ_{ng}. We can write the expression for only one nonzero polarizability tensor component in the laboratory coordinate system related to the individual nanoparticle using expressions (13) and (16), and introducing the definition $\Delta\omega_{ng}=\omega-\omega_{ng}$:

$$\chi(\omega, Q_1, Q_2) = N\sum_{n,g} \left[\frac{A_{ng}\Delta\omega_{ng}}{\Box\left(\Delta\omega_{ng}^2 + \Gamma_{ng}^2\right)} - i\frac{A_{ng}\Gamma_{ng}}{\Box\left(\Delta\omega_{ng}^2 + \Gamma_{ng}^2\right)} \right] \Delta\rho_{ng} \tag{17}$$

The next definition is included into the expression (17)

$$A_{ng}(Q_1, Q_2) = \frac{1}{3}\mid p_{ng} \mid^2 + Q_1\left(\mid p_1^{ng} \mid^2 - \mid p_2^{ng} \mid^2\right) + Q_2\left(\mid p_3^{ng} \mid^2 - \mid p_2^{ng} \mid^2\right) \tag{18}$$

The quantity A_{ng} is proportional to the squared magnitude of the dipole moment of transitions from the state <n| to the state |g> provided certain optical radiation intensities, frequencies and specified parameters of the nanocomposite matrix. The population difference induced by radiation between the states <n| and |g> is a function of the incident radiation intensity. Using a two-level system approximation [52], this difference is

$$\Delta\rho_{ng}(I) = \left(1 - \frac{I/I_S}{\Delta\omega_{ng}^2 + \Gamma_{ng}^2\left(1 + I/I_S\right)}\Gamma_{ng}^2\right) \Delta\rho_{ng}^0 \tag{19}$$

where $\Delta\varrho^0{}_{ng}$ is the thermal-equilibrium difference and I_s is the intensity of saturation, when the $\Delta\varrho/2$ carriers are in the upper energy level. Separating the real and imaginary parts of the polarizability tensor component (17) and taking into account the expression (19) we introduce the definitions

$$P = \int \sin^2\theta d\Omega$$

$$B_{ng}(\omega, T) = \frac{I\Big/I_S}{\Delta\omega_{ng}^2 + \Gamma_{ng}^2\left(1 + I\Big/I_S\right)}\Gamma_{ng}^2 \tag{20}$$

we obtain the integrated scattering and absorption cross sections of the united volume of the HDN in a single scattering approximation:

$$\sigma_a(\omega, a, I) = \frac{4\pi\omega N}{c\Box} \sum_{n,g} \frac{A_{ng}\Gamma_{ng}\Delta\rho_{ng}^0}{\Delta\omega_{ng}^0 + \Gamma_{ng}^2}(1 - B_{ng}) \tag{21}$$

$$\sigma_S(\omega, a, I) =$$

$$\frac{\omega^4 P N^2}{c^4\Box^2} \sum_{n,g}\sum_{k,l}\left\{A_{ng}A_{kl}\frac{(\Delta\omega_{ng}\Delta\omega_{kl} + \Gamma_{ng}\Gamma_{kl})\Delta\rho_{ng}^0\Delta\rho_{kl}^0}{(\Delta\omega_{ng}^2 + \Gamma_{ng}^2)(\Delta\omega_{kl}^2 + \Gamma_{kl}^2)}\ (1 - B_{ng})(1 - B_{kl})\right\} \tag{22}$$

The dependence of the cross sections on the nanoparticle dimensions can be found by knowing the function of the relation between A_{ng} and the nanoparticle dimension. Given $\alpha<\alpha_c$ the dipole moment of the nanoparticle is proportional to its dimension. Therefore, as follows from expression (18), we can separate out the dependence of A_{ng} on the nanoparticle dimension as $A_{ng}=S_{ng}(I)a^2$. Here, $S_{ng}(I)$ is a function of the radiation intensity and depends on the nanoparticle shape.

Let us estimate the ratio between the scattering and absorption cross sections. We assume that transitions occur from only one level $<n|$ and the width of the excited level Γ_g has a low-dependence on g. Taking into account that the frequencies ω and ω_g are of one magnitude order and the thermal equilibrium difference between the states is close to unity, and following the expressions (21) and (22), we obtain the next

$$\frac{\sigma_S(\omega, a)}{\sigma_a(\omega, a)} \approx \frac{NP\omega^3 a^2}{4\pi c^3\Box\Gamma}\sum_{n,g}\{S_{ng}(I)\ (1 - B_{ng}(I))\} \tag{23}$$

The quantity of σ_S/σ_a does not exceed $N*10^{-9}$ in any intensity region provided the nanoparticle dimensions $\alpha=(10;100)$nm in the frequency range $(10^{13};10^{16})$Hz and $\Gamma=10^9$Hz. Given $N*10^{-9}<1$, the scattering cross section can be omitted from the expression for the extinction coefficient.

We can define $S_{ng}(I)=c_{ng}I$ and follow the expression (21) within radiation intensities region $I/I_s \ll 1$, we obtain the next

$$\sigma_a(\omega, a, I) \approx \frac{4\pi\omega N}{c\square} a^2 I \sum_{n,g} c_{ng} \frac{\Gamma_{ng}}{(\Delta\omega_{ng}^2 + \Gamma_{ng}^2)^2} \Delta\rho_{ng}^0 \qquad (24)$$

The absorption cross section reaches a maximum at some intensity $I=I_p$ under increasing radiation intensity, that follows from equation (21). It corresponds to the complete nanoparticle's orientation along the electric vector of the optical field and to the maximum value of $A_{ng}(I)$. This effect is responsible for a sharp enhancement of radiation absorption by the HDN unit volume. A further intensity increasing yields a noticeable increase of $B_{ng}(\omega,I)$ and decrease of the absorption cross section at a constant value of $A_{ng}(I)$. Given $I \gg I_s$ the value of $B_{ng}(\omega,I)$ becomes approximately equal to unity resulting in an increase of the HDN transmittance. In this case, the absorption cross section can be written as

$$\sigma_a(\omega, a, I) \approx \frac{4\pi\omega N}{c\square} a^2 \sum_{n,g} S_{ng} \frac{I_s}{I} \frac{1}{\Gamma_{ng}} \Delta\rho_{ng}^0 \qquad (25)$$

i.e., it is inverse proportional to the radiation intensity.

The broad optical absorption bands are manifested in the electronic structure of dielectric nanoparticles and absent in the corresponding bulk sample. In addition, the allowed electron energy sub-band (excitons, impurities, etc.) with the width $\Delta\omega_1$ lying in the band gap and adjoining the conduction band bottom, as well as the size-quantization levels (minibands) with the width $\Delta\omega_2$ in the conduction band, are typical for the electronic structure of HDN electrons. Taking into account the electronic structure of the nanoparticle, we substitute the summation over $|g>$ states with integration from $(\omega_n - \Delta\omega_1)$ to $(\omega_n + \Delta\omega_2)$ with state densities of exciton g_1 and quantum-size g_2 levels, respectively. Let us choose one of the absorption bands as an example. Changing the summation by the integration over the frequency in (21) and introducing the definitions

$$\Delta\omega_n = \omega - \omega_n$$
$$F(I) = \sqrt{\frac{I_s}{I + I_s}} \qquad (26)$$

we obtain the expression for the absorption cross section of light within the absorption band:

$$\sigma_a = \frac{4\pi\omega N a^2}{c\square} F(I) \left[g_1 S_1 \arctan\left(\frac{\Delta\omega_1 \Gamma_n F(I)}{\Gamma_n^2 + F^2(I)\Delta\omega_n(\Delta\omega_n + \Delta\omega_1)} \right) \right.$$
$$\left. + g_2 S_2 \arctan\left(\frac{\Delta\omega_2 \Gamma_n F(I)}{\Gamma_n^2 + F^2(I)\Delta\omega_n(\Delta\omega_n - \Delta\omega_2)} \right) \right]$$

(27)

The quantities S_1 and S_2 are defined as the average form factors of nanoparticles $S_{ng}(I)$ for transitions to the upper and lower energy bands, respectively.

We may obtain the expression for the HDN optical transmittance in the absorption band with the central frequency ω_n from expression (7) and (27):

$$T(\omega, N, I) \approx \exp\left\{ -L \frac{4\pi\omega N}{c\square} DF(I) \right\}$$

(28)

where

$$D = a^2 g_1 S_1 \arctan\left(\frac{\Delta\omega_1 \Gamma_n F(I)}{\Gamma_n^2 + F^2(I)\Delta\omega_n(\Delta\omega_n + \Delta\omega_1)} \right)$$
$$+ a^2 g_2 S_2 \arctan\left(\frac{\Delta\omega_2 \Gamma_n F(I)}{\Gamma_n^2 + F^2(I)\Delta\omega_n(\Delta\omega_n - \Delta\omega_2)} \right)$$

(29)

3.3. The Theoretical Outputs and Discussion

It follows from expression (28) that the optical transmittance of the HDN essentially depends on the laser radiation intensity I (Figure 13). This dependence exhibits a minimum I_p corresponding to the lowest light transmittance of the HDN. As intensity is changed near I_p, we can see the effect of limitation of low-intensity radiation. The insert in Figure 13 is the experimental result obtained from Figure 10B. Since $I_{out}=I_{in}e^{-\alpha L}$ and $T=I_{out}/I_{in}$ provided low reflection and absorption, we suppose $T=e^{-\alpha L}$ and use data of α from Figure 10B. The theoretical and experimental results are in good agreement.

Curves from Figure 14 point out the basic features of the dependences of the HDN transmittance on the radiation wavelength. In the general case, the transmittance spectrum is asymmetric, since there is the difference between $\Delta\omega_1$ and $\Delta\omega_2$. The insert in this figure is the experimental spectrum of the HDN based on SiO_2 nanoparticles (Figure 4B). The behaviour of the experimental curve reflects the features of the theoretical one.

The largest dipole moment is induced at the central frequency ω_n in the absorption band. The expression for D at the central frequency of the absorption band ($\omega=\omega_n$, $\Delta\omega_n=0$) is given by

$$D = a^2 g S \left\{ \arctan\left(\frac{\Delta\omega_1 F(I)}{\Gamma_n} \right) + \arctan\left(\frac{\Delta\omega_2 F(I)}{\Gamma_n} \right) \right\}$$

(30)

Therefore, the HDN transmittance near the central frequency can be written as

$$T(I) = \exp\left\{- L\ \frac{4\pi\omega_n N}{c\Box} a^2 g S\left[\arctan\left(\frac{\Delta\omega_1 F(I)}{\Gamma_n}\right) + \arctan\left(\frac{\Delta\omega_2 F(I)}{\Gamma_n}\right)\right] F(I)\right\}$$ (31)

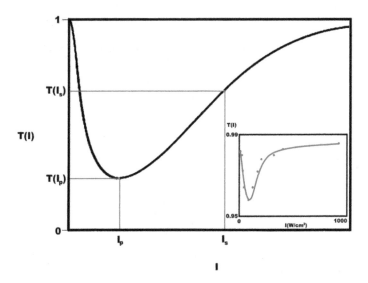

Figure 13. The theoretical dependence of the HDN transmittance on the intensity of input radiation. The insert is the experimental dependence of the HDN transmittance according with Figure 10B.

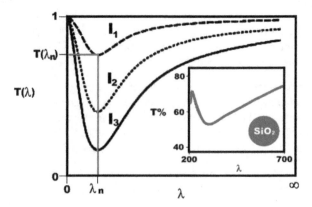

Figure 14. The theoretical dependence of the HDN transmittance on the wavelength of input radiation ($I_1 > I_2 > I_3$). The insert is the experimental spectrum of the HDN based on SiO_2 nanoparticles (Figure 4B).

As one can see from expression (31), the depth of the absorption band in the transmittance spectrum depends on the radiation intensity and the nanoparticle dimension. The orientation of nanoparticles along the vector E requires high radiation intensities provided a solid HDN. In the case of a liquid matrix, this situation corresponds to the range of intensities $I \gg I_p$. Here we may assume that all particles are oriented along the direction of the vector E of the external optical field, so the order parameters are constant and S_n are independent of the intensity. This indicates that the behaviour of the transmittance of solid and liquid matrices is similar. Therefore, the transmittance at the central frequency is

$$T(I) = \exp\left\{ -L \ \frac{4\pi\omega_n N}{c\Box\Gamma_n} a^2 (gS_n)_{g=n} (\Delta\omega_1 + \Delta\omega_2) F^2(I) \right\} \tag{32}$$

Expression (32) exponentially approaches to unity rapidly when the radiation intensity is increased and nanoparticle dimension, the summation $(\Delta\omega_1 + \Delta\omega_2)$ and multiplication of gS, become larger.

Apart from transmittance, scattering and spectral properties of the HDN, the theory can describe the behaviour of light refraction in the HDN (Figures 15,16). Using expressions (16) and (19), we can obtain the theoretical dependence of the refractive index on the intensity. Since the value of $\Delta n(I)$ is negligible, the medium refractive index can be written as follows:

$$n(I, \lambda) \approx n_0 + \frac{2\pi\chi_{zz}(I, \lambda)}{n_0} \tag{33}$$

where n_0 is the HDN refractive index in the absence of radiation and χ_{zz} is defined by expression (13).

Since χ_{zz} is determined by α_{ij}, so we may simplify equation (16) to carry out the integration over frequency, herewith, to assume $\Gamma_{ng} = \Gamma_n$ and the state density g_1 (g_2) and the values of Q_1 (Q_2) are independent of the frequency ω. Picking out the refraction real part from the resulting expression, we obtain [9,45]

$$n(I, \omega) = n_0 + \frac{\Box}{2} \sum_n \left| A_{ng}(Q_1, Q_2) \Delta\rho^0 \left| g_1 \ln \frac{(\omega - (\omega_n - \Delta\omega_1))^2 + \Gamma_n^2 \left(1 + \frac{I}{I_S} \right)}{(\omega - \omega_n)^2 + \Gamma_n^2 \left(1 + \frac{I}{I_S} \right)} + g_2 \ln \frac{(\omega - \omega_n)^2 + \Gamma_n^2 \left(1 + \frac{I}{I_S} \right)}{(\omega - (\omega_n + \Delta\omega_2))^2 + \Gamma_n^2 \left(1 + \frac{I}{I_S} \right)} \right| \right| \tag{34}$$

where $A_{ng}(Q_1, Q_2)$ is determined by expression (18).

Equation (34) indicates that the term A_{ng} does not vanish in the case of propagation of unpolarized light through the medium (Q=0) and the nonlinear response of the dielectric nanocomposite is not equal to zero even in case of a solid matrix.

The important conclusion from equations (18) and (34) is that the modulus of photo-induced electric dipole moments $|p_{ng}|$ mainly defines the magnitude of nonlinear optical response under continuous low-intensity radiation. In general, the orientation parametersand dipole moment modulus reach their maxima with the increase of input power, however, this increase diminishes the population difference; hence the change of Δn tends to zero. It is these two competing processes that define the nonlinear features of the HDN refractive index.

The numerical simulation of the change in the HDN refractive index was carried out using equation (34). Since the photon energy is less than nanoparticle band gap, so, the dipole transition of electrons to the exciton state is most probable ($g_2=0$). In the case of the low-intensity continuous radiation and low concentration of nanoparticles, equation (34)can be rewritten as follows [9]:

$$\Delta n(I, \omega) \approx A(I)\ln \frac{(\omega - (\omega_n - \Delta\omega_1))^2 + \Gamma_n^2\left(1 + \frac{I}{I_s}\right)}{(\omega - \omega_n)^2 + \Gamma_n^2\left(1 + \frac{I}{I_s}\right)} \tag{35}$$

where the factor A(I) defines the dependence of A_{ng} on the radiation intensity as follows:

$$A(I) = A_0 \; \Delta\rho^0\left(1 - e^{-al}\right) \tag{36}$$

This dependence takes into account the magnitude of A_{ng}, which varies from zero to its maximum value with the increase of the external radiation intensity. The theoretical curves (Figures 15,16, solid lines) of the dependence Δn on the radiation intensity have been constructed by means of equation (35) for the HDN based on Al_2O_3 and SiO_2 nanoparticles, irradiating by green and violet radiation. The parameters for good approximation were calculated according to the next considerations: Γ was taken from T spectrum (Figure 4); I_s, α and $A_0\Delta\rho^0$ were calculated by means of the three-equation system (35) with known parameters I and Δn (Figures 9 A-B, 10A, 11A, dotted curves).

The lack of nonlinear refraction and absorption of low-intensity continuous visible laser radiation in the HDN based on the nanoparticles of narrow-band dielectrics is caused by the absence of absorption band in the used frequency range (200;700)nm. So, in order for the nonlinear optical properties of such HDN to be observable we must use high-energy pulsed radiation or change the input radiation frequency. On the one hand, if we use the pulsed radiation we may get the typical nonlinearity of the nanocomposites, caused by nonlinear behaviourof excitons near the edge of fundamental absorption (Figure 6). The high energy is required, since there is the small dipole moment of electron transition to exciton states. On the other hand, if we change the input radiation frequency it is possible to find the defect energy levels in the HDN infrared spectrum. So, the nontypical nonlinearity can take place under low-intensity infrared radiation.

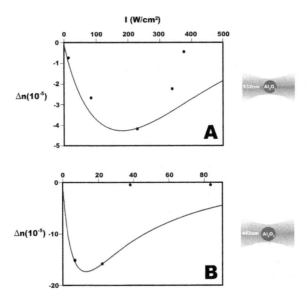

Figure 15. Theoretical curves of dependence of refraction index of green (A) and violet (B) radiation on its intensity in the HDN based on Al_2O_3 nanoparticles (dotted curves are the experimental results from Figure 9).

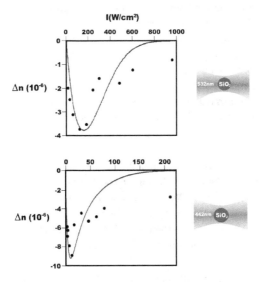

Figure 16. Theoretical curves of dependence of refraction index of green (A) and violet (B) radiation on its intensity in the HDN based on SiO_2 nanoparticles (dotted curves are the experimental results from Figures 9A and 10A).

4. Conclusion

The experimental study of changes of optical characteristics of the dielectric nanostructures based on Al_2O_3, SiO_2, TiO_2, ZnO nanoparticles and theoretical description of these characteristics allows estimating the conditions of observing the low-threshold optical nonlinearity under low-intensity optical fields.

The ability to observe this nonlinearity is directly connected with the peculiarities of the energy spectrum of nanoparticle charge carriers. Because of the wide band gap of the bulk dielectric material, it is not possible to excite electron transitions to the conduction band by a visible light. The energy spectrum of nanoparticle electrons is of a different structure: the band gap has defect levels containing a lot of electrons due to a high density of crystal defects on the nanoparticle's surface; the small size and shape of nanoparticle leads to strong broadening of the band of high-density exciton states from the bottom of the conduction band up to defect levels. The existence of an absorption band in visible light spectrum is observed only for nanoparticles of broad-band dielectrics (Al_2O_3, SiO_2). The absorption band in the energy spectrum of electrons of narrow-band dielectric (TiO_2, ZnO) nanoparticles is not manifested in a visible light spectrum, however, it can be manifested within the infrared region and adjoins the bottom of the conduction band.

Comparing the experimental and theoretical results we conclude that the low-threshold nonlinearity of the HDN optical parameters (Δn, $\Delta\alpha$ and the scattering cross section) is caused, mainly, by transitions of electrons from defect levels to exciton states and, hence, photo excitation of electric dipole moments. However, experiments have shown that the nonlinear behaviour of the HDN optical parameters takes place when the matrix permittivity ε_{stat} is less than that of the nanoparticles (e.g., oil permittivity). Otherwise, the positive polarization charges, concentrated along the nanoparticle's inner surface, destroy the defect states. This explains the absence of nonlinear optical properties in the HDN based on water and alcohol matrices.

In view of the effect of giant oscillator strength, the magnitude of the photo excited dipole moment is enormous. It is the great value of the oscillator's strength for electron transition to the exciton states that is responsible for the low-threshold of the nonlinearity. As it follows from the theory, the dipole moment orientation along the external E field makes a minor contribution to the nonlinearity, therefore, this response can also be observed under pulsed and unpolarized laser radiation in solid matrices.

In addition, a qualitative agreement between experimental and theoretical results was also obtained and the proposed theory model of optical nonlinearity can be applied to explain the number of phenomena in physics of nanoscale dielectrics, e.g., proteins and blood bodies [53].

Acknowledgements

This work was supported by RFBR Grant No. 11-02-98514 r_vostok_a and FEB RAS Grant Nos. 12-I- OFN-05, 12-I- OFN-04, 12-I-P24-05, 12-II-UO-02-002.

Author details

Vladimir Dzyuba[1,2*], Yurii Kulchin[1,2] and Valentin Milichko[1,2]

*Address all correspondence to: vdzyuba@iacp.dvo.ru

1 Institute of Automation and Control Processes of Russian Academy of Science, Vladivostok, Russia

2 Far Eastern Federal University, Vladivostok, Russia

References

[1] Alivisatos, A. P., Harris, A. L., Levinos, N. J., Steigerwald, M. L., & Brus, L. E. (1988). Electronic states of semiconductor clusters: Homogeneous and inhomogeneous broadening of the optical spectrum. *The Journal of Chemical Physics*, 89(7), 4001-4011.

[2] Bawendi, M. G., Carroll, P. J., Wilson, William L., & Brus, L. E. (1992). Luminescence properties of CdSe quantum crystallites: Resonance between interior and surface localized states. *The Journal of Chemical Physics*, 96(2), 946-954.

[3] Bang, Jin Ho., & Kamat, Prashant V. (2009). Quantum Dot Sensitized Solar Cells. A Tale of Two Semiconductor Nanocrystals: CdSe and CdTe. *ACS Nano*, 3(6), 1467-1476.

[4] McGuire, John A., Sykora, Milan., Robel, Istva´n., Padilha, Lazaro A., Joo, Jin., Pietryga, Jeffrey. M., & Klimov, Victor. I. (2010). Spectroscopic Signatures of Photocharging due to Hot-Carrier Transfer in Solutions of Semiconductor Nanocrystals under Low-Intensity Ultraviolet Excitation. *ACS Nano*, 4(10), 6087-6097.

[5] Ivanov, S. A., & Achermann, M. (2010). Spectral and Dynamic Properties of Excitons and Biexcitons in Type-II Semiconductor Nanocrystals. *ACS Nano*, 4(10), 5994-6000.

[6] Hashimoto, Tadanori., Yamamoto, Tsuyoshi., Kato, Tomohiro., Nasu, Hiroyuki., & Kamiya, Kanichi. (2001). Z-scan analyses for PbO-containing glass with large optical nonlinearity. *Journal of Applied Physics*, 90(2), 533-537.

[7] Anderson, Mark S. (2003). Enhanced infrared absorption with dielectric nanoparticles. *Applied Physics Letters*, 83(14), 2964-2966.

[8]   Miheev, O. P., & Sidorov, A. I. (2004). Optical nonlinearity of nanoparticles of wide-gap semiconductors and insulators in visible and near infrared spectral region. *Technical Physics*, 74(6), 77-82.

[9]   Dzyuba, Vladimir., Milichko, Valentin., & Kulchin, Yurii. (2011). Nontypical photoinduced optical nonlinearity of dielectric nanostructures. *Journal of Nanophotonics*, 5, 053528, 1-13.

[10] Yu, Baolong, Zhu, Congshan, & Gan, Fuxi. (1997). Optical nonlinearity of Bi_2O_3 nanoparticles studied by Z-scan technique. *Journal of Applied Physics*, 82(9), 4532-4537.

[11] Major, A., Yoshino, F., Aitchison, J. S., & Smith, P. W. E. (2004). Ultrafast nonresonant third-order optical nonlinearities in ZnSe for photonic switching at telecom wavelengths. *Applied Physics Letters*, 85(20), 4606-4608.

[12] He, J., Ji, W., Ma, G. H., Tang, S. H., Elim, H. I., Sun, W. X., Zhang, Z. H., & Chin, W. S. (2004). Excitonic nonlinear absorption in CdS nanocrystals studied using Z-scan technique. *Journal of Applied Physics*, 95(11), 6381-6386.

[13] Larciprete, M. C., Ostuni, R., Belardini, A., Alonzo, M., Leahu, G., Fazio, E., Sibilia, C., & Bertolotti, M. (2007). Nonlinear optical absorption of zinc-phthalocyanines in polymeric matrix. *Photonics and Nanostructures - Fundamentals and Applications*, 5, 73-78.

[14] Ganeev, R. A., Suzuki, M., Baba, M., Ichihara, M., & Kuroda, H. (2008). Low- and high-order nonlinear optical properties of $BaTiO3$ and $SrTiO3$ nanoparticles. *Journal of Optical Society of America B*, 25(3), 325-333.

[15] Ganeev, R. A., Zakirov, A. S., Boltaev, G. S., Tugushev, R. I., Usmanov, T., Khabibullaev, P. K., Kang, T. W., & Saidov, A. A. (2003). Structural, optical, and nonlinear optical absorption/refraction studies of the manganese nanoparticles prepared by laser ablation in ethanol. *Optical Quantum Electronics*, 35, 419-423.

[16] Ganeev, R. A., Ryasnyansky, A. I., Tugushev, R. I., & Usmanov, T. (2003). Investigation of nonlinear refraction and nonlinear absorption of semiconductor nanoparticle solutions prepared by laser ablation. *Journal of Optics A*, 5, 409-417.

[17] Ganeev, R. A., & Usmanov, T. (2007). Nonlinear-optical parameters of various media. *IOP Quantum Electron*, 37(7), 605-622.

[18] Kulchin, Yu. N., Shcherbakov, A. V., Dzyuba, V. P., & Voznesenskiy, S. S. (2009). Interaction of collinear light beams with different wavelengths in a heterogeneous liquid-phase nanocomposite. *Technical Physics Letters*, 35(7), 640-642.

[19] Chen, Fang-Chung., Chu, Chih-Wei., He, Jun., Yang, Yang., & Lin, Jen-Lien. (2004). Organic thin-film transistors with nanocomposite dielectric gate insulator. *Applied Physics Letters*, 85(15), 3295-3297.

[20] Schrier, Joshua., Demchenko, Denis O., Wang, Lin Wang., & Alivisatos, A. Paul. (2007). Optical Properties of ZnO/ZnS and ZnO/ZnTe Heterostructures for Photovoltaic Applications. *Nano Letters*, 7(8), 2377-2382.

[21] Dutta, Kousik., & De, S. K. (2007). Electrical conductivity and dielectric properties of SiO2 nanoparticles dispersed in conducting polymer matrix. *Journal of Nanoparticle Research*, 9, 631-638.

[22] Chu, Daobao, Yuan, Ximei, Qin, Guoxu, Xu, Mai, Zheng, Peng, Lu, Jia, & Zha, Long-wu. (2008). Efficient carbon-doped nanostructured TiO2 (anatase) film for photoelectrochemical solar cells. *Journal of Nanoparticle Research*, 10, 357-363.

[23] Liu, Li., Wang, Ning., Cao, Xia., & Guo, Lin. (2010). Direct Electrochemistry of Cytochrome c at a Hierarchically Nanostructured TiO2 Quantum Electrode. *Nano Research*, 3-369.

[24] Hensel, Jennifer., Wang, Gongming., Li, Yat., & Zhang, Jin. Z. (2010). Synergistic Effect of CdSe Quantum Dot Sensitization and Nitrogen Doping of TiO2 Nanostructures for Photoelectrochemical Solar Hydrogen Generation. *Nano Letters*, 10, 478-483.

[25] Pellegrini, Giovanni., Mattei, Giovanni., & Mazzoldi, Paolo. (2009). Light Extraction with Dielectric Nanoantenna Arrays. *ACS Nano*, 3(9), 2715-2721.

[26] Redel, Engelbert., Mirtchev, Peter., Huai, Chen., Petrov, Srebri., & Ozin, Geoffrey. A. (2011). Nanoparticle Films and Photonic Crystal Multilayers from Colloidally Stable, Size-Controllable Zinc and Iron Oxide Nanoparticles. *ACS Nano*, 5(4), 2861-2869.

[27] Chen, F. C., Chuang, C. S., Lin, Y. S., Kung, L. J., Chen, T. H., & Shien, D. H. P. (2006). Low voltage organic thin film transistors with polymeric nanocomposite dielectrics *Organic Electronics*, 7, 435-439.

[28] Park, Young-Shin., Cook, Andrew K., & Wang, Hailin. (2006). Cavity QED with Diamond Nanocrystals and Silica Microspheres. *Nano Letters*, 6(9), 2075-2079.

[29] Han, Jiaguang., Woo, Boon Kuan., Chen, Wei., Sang, Mei., Lu, Xinchao., & Zhang, Weili. (2008). Terahertz dielectric properties of MgO nanocristals. *The Journal of Physical Chemistry C*, 112, 17512-17516.

[30] Balasubramanian, Balamurugan., Kraemer, Kristin L., Reding, Nicholas A., Skomski, Ralph., Ducharme, Stephen., & Sellmyer, David J. (2010). Synthesis of Monodisperse TiO2-Paraffin Core-Shell Nanoparticles for Improved Dielectric Properties. *ACS Nano*, 4(4), 1893-1900.

[31] Landes, C., Braun, M., Burda, C., & El -Sayed, M. A. (2001). Observation of Large Changes in the Band Gap Absorption Energy of Small CdSe Nanoparticles Induced by the Adsorption of a Strong Hole Acceptor. *Nano Letters*, 1(11), 667-670.

[32] Kulchin, Yu. N., Shcherbakov, A. V., Dzyuba, V. P., Voznesenskii, S. S., & Mikaelyan, G. T. (2008). Nonlinear-optical properties of heterogeneous liquid nanophase composites based on high-energy-gap Al2O3 nanoparticles. *IOP Quantum Electronics*, 38(2), 154-158.

[33] Ho, Ching-Hwa., Chan, Ching-Hsiang., Tien, Li-Chia., & Huang, Ying-Sheng. (2011). Direct Optical Observation of Band-Edge Excitons, Band Gap, and Fermi Level in Degenerate Semiconducting Oxide Nanowires In2O3. *Journal of Physical Chemistry C*, 115, 25088-25096.

[34] Mu, R., Tung, Y. S., Ueda, A., & Henderson, D. O. (1996). Chemical and Size Characterization of Layered Lead Iodide Quantum Dots via Optical Spectroscopy and Atomic Force Microscopy. *Journal of Physical Chemistry*, 100, 19927-19932.

[35] Li, Jingbo., & Wang, Lin-Wang. (2003). Shape Effects on Electronic States of Nanocrystals. *Nano Letters*, 3(10), 1357-1363.

[36] Al-Hilli, S. M., & Willander, M. (2006). Optical properties of zinc oxide nano-particles embedded in dielectric medium for UV region: Numerical simulation. *Journal of Nanoparticle Research*, 8, 79-97.

[37] Stroyuk, Oleksandr. L., Dzhagan, Volodymyr. M., Shvalagin, Vitaliy. V., & Kuchmiy, Stepan Ya. (2010). Size-Dependent Optical Properties of Colloidal ZnO Nanoparticles Charged by Photoexcitation. *The Journal of Physical Chemistry C*, 114, 220-225.

[38] Azpiroz, Jon M., Mosconi, Edoardo., & De Angelis, Filippo. (2011). Modeling ZnS and ZnO Nanostructures: Structural, Electronic, and Optical Properties. *The Journal of Physical Chemistry C*, 115, 25219-25226.

[39] Pokutnyi, S. I. (2006). Optical absorption and scattering at one-particle states of charge carriers in semiconductor quantum dots. *Semiconductor*, 40(2), 217-223.

[40] Pokutnyi, S. I. (2007). Excition states in semiconductor quantum dots in the modified effective mass approximation. *Semiconductor*, 41(11), 1323-1328.

[41] Sinha, Sucharita., Ray, Alok., & Dasgupta, K. (2000). Solvent dependent nonlinear refraction in organic dye solution. *Journal of Applied Physics*, 87(7), 3222-3226.

[42] Mosconi, Edoardo., Selloni, Annabella., & De Angelis, Filippo. (2012). Solvent Effects on the Adsorption Geometry and Electronic Structure of Dye-Sensitized TiO2: A First-Principles Investigation. *Journal of Physical Chemistry C*, 116, 5932-5940.

[43] Dzyuba, V. P., Krasnok, A. E., Kulchin, Yu. N., & Dzyuba, I. V. (2011). A model of nonlinear optical transmittance for insulator nanocomposites. *Semiconductor*, 45(3), 295-301.

[44] Dzyuba, V. P., Krasnok, A. E., & Kulchin, Yu. N. (2010). Nonlinear refractive index of dielectric nanocomposites in weak optical fields. *Technical Physics Letters*, 36(11), 973-977.

[45] Kulchin, Yu. N., Dzyuba, V. P., & Voznesenskiy, S. S. (2011). Threshold optical nonlinearity of dielectric nanocomposite. In: Boreddy Reddy (ed.), *Advances in Diverse Industrial Applications of Nanocomposites*, Rijeka, InTech, 261-288.

[46] Wakaki, M., Kudo, K., & Shibuya, T. (2007). Physical Properties and Data of Optical Materials. CRC Press Taylor and Francis Group. New York. 10.1201/9781420015508

[47] Said, A. A., Sheik-Bahae, M., Hagan, D. J., Wei, T. H., Wang, J., Young, J., & Van Stry-land, E. W. (1991). Determination of bound-electronic and free-carrier nonlinearities in ZnSe, GaAs, CdTe and ZnTe. *Journal of Optical Society of America B*, 9(3), 405-414.

[48] Van Stryland, E. W., & Sheik-Bahae, M. (1998). Z-scan measurement of optical nonli-nearities. In: Kuzyk M. G., DirkC. W. (eds), *Characterization Techniques and Tabulation for Organic Nonlinear Materials*, Marcel Dekker Inc., New York, 655-692.

[49] Pokitnyi, S. I., & Efremov, N. A. (1991). Theory of Macroscopic Local Single-Particle Charge States in Quasi-Zero-Dimensional Structures Surface Local States. *Phys. Sta-tus Solidi B*, 165, 109-118.

[50] Bohren, C., & Huffman, D. (1983). *Absorbtion and Scattering of Light by Small Particles*, Wiley, New York.

[51] Landau, L. D., & Lifshitz, E. M. (1984). Course of Theoretical Physics, Volume 8: Elec-trodynamics of Continuous Media. Pergamon, New York.

[52] Shen, Y. R. (1984). The Principles of Nonlinear Optic. Wiley, Hoboken, NJ.

[53] Kulchin, Yu. N., Dzyuba, V. P., & Milichko, V. A. (2010). Optical Nonlinearity of a Biological Liquid Nanocomposite. *Pacific Science Review*, 12(1), 4-7.

Chapter 6

Effect of Nano-TiN on Mechanical Behavior of Si$_3$N$_4$ Based Nanocomposites by Spark Plasma Sintering (SPS)

Jow-Lay Huang and Pramoda K. Nayak

Additional information is available at the end of the chapter

1. Introduction

Ceramic nanocomposites are often defined as a ceramic matrix reinforced with submicron/ nano sized particles of a secondary phase. The advantages of these nanocomposites include: improved mechanical properties, surface properties, high thermal stability and superior thermal conductivity. It is very fascinating/interesting for the researchers to synthesize these composites as the incorporation of few percent nanosized particles changes the materials property substantially. Niihara et al.,[35], [36] have reported that the mechanical properties of ceramics can be improved significantly by dispersing nanometer-sized ceramic particles into ceramic matrix grains or grain boundaries. According to their observation, 5 vol% of silicon carbide nanoparticles into alumina matrix increases the room temperature strength from 350 MPa to approximately 1 GPa. Other strength improvements through similar approaches have been observed in alumina-silicon nitride, magnesia-silicon carbide, and silicon nitride-silicon carbide composite systems.

Apart from the basic mechanical properties such as such as micro hardness, facture strength, and facture toughness [9; 23; 44], nanocomposites also exhibit electro conductive, wear resistance, creep resistance and high temperature performance [10; 24; 37; 38, 39] However, the degree of improvement in these properties is dependent on the type of composite system involved.

1.1. Novel Synthesis of Ceramic nanocomposite

Chemical Vapor Deposition (CVD) is a very preferable method to disperse the nano-sized second phases into the matrix grains or at the grain boundaries [33]. However, the CVD process is not applicable to fabricate the large and complex shaped component for the mass production and also it is very expensive. Processing route is another technique to prepare ceramic nanocomposites. Following the initial work of [34], several research groups have tried to synthesize the nanocomposites using different processing route such as conventional powder processing [6; 8], hot press sintering [25; 47] sol-gel processing [30; 50] and polymer processing [5; 16]. The ceramic nanocomposites can be synthesized using microwave plasma [48; 49]. The main advantage of this technique is that the reaction product does not form hard agglomerates because of the specific conditions during synthesis.

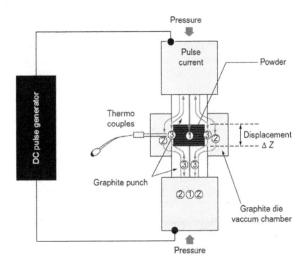

Figure 1. Schematics of Spark Plasma Synthesis (SPS) process

Recently developed Spark plasma sintering (SPS) is a novel sintering technique that uses the idea of pressure driven powder consolidation under pulsed direct electric current passing through a sample compressed in a graphite matrix. It is also known as the field assisted sintering technique or pulse electric current sintering. This newly developed sintering technique is regarded as an energy-saving technology due to the short process time and fewer processing steps. This technique was first described by Raichenko, 1987 and the key characteristics of this SPS are given as follows:

i.   The generation of local electric discharge plasma and its effect on the material.

ii.   The combined effect of external fields, such as a force field and electric field, on the densification and phase formation in a particulate system.

iii.   The influence of electric current in the near surface layers of conductors and semi-conductors (the so-called "skin-effect").

iv.   The rapid and nonuniform heating/cooling throughout the sample, causing large temperature gradients.

The schematic of SPS process is shown in Fig.1 [41].

1.2. Advantages of SPS over other synthesis method

The most impressive advantage of SPS is its applicability to sinter materials of various types of chemical bonding and electric conductivity. Novel materials have been prepared from powders of ceramic dielectrics, conductors, semiconductors, amorphous alloys, and, sometimes, polymers. The traditional driving force involved in commonly used consolidation techniques such as solid state sintering and hot press sintering are: surface tension, external pressure, chemical potential due to the gradient of surface curvature, concentration gradient in multicomponent system etc. In SPS technique, the additional driving forces include electromechanical stress, high local temperature gradients creating thermal stresses intensifying thermal diffusion, and dislocation creep. These additional driving forces are responsible for much faster transport mechanism, that accelerates rapid sintering, which is observed in SPS.

1.3. TiN/Si_3N_4 Ceramic Nanocomposites

Si_3N_4 ceramics are regarded as one of the important high temperature structural materials. These ceramics have attracted much attention due to their good mechanical and chemical properties and due to their reliability at room and elevated temperatures [14; 19]. High strength and high-toughness Si_3N_4 matrix composites such as whisker-reinforced of particulate-reinforced ceramics, have been developed to improve the mechanical reliability of Si_3N_4 ceramics [2; 13]. However, these composites are extremely hard and machining using conventionals tools is difficult, which limits the widespread application of these materials in many fields. If, sintered Si_3N_4 bodies can be made electro conductive, electrical discharge machining (EDM) technique can be applied to manufacture complex components [32]. It has been reported that introduction of electro conductive second phase can improve the mechanical properties and electroconductivity of Si_3N_4 ceramics [14; 18; 43].

TiN exhibits a number of desirable properties, including high hardness, good chemical durability, high electrical conductivity and is a popular second phase additive due to its good compatibility with Si_3N_4. It often incorporated into the β-Si_3N_4 matrix as cutting-tool materials [4; 14; 17; 28]. There are two advantages to the Si_3N_4 based composites. First of all, the good physical properties of TiN, such as high melting point, hardness, strength and chemical stability, as well as its good erosion and corrosion resistance, enable it to be an excellent toughening material [7; 28; 51]. Secondly, the electrical resistance of Si_3N_4 can be substantially decreased, which consequently makes electric-discharge machining possible [17; 21].

2. Spark Plasma Sintering of TiN/Si_3N_4 nanocomposite

As described in the previous section, SPS is a newly developed sintering technique and it is beneficial to consolidate Si_3N_4 based nanocomposites in a short time. Some researchers have already reported that TiN/Si_3N_4 based nanocomposites with excellent mechanical properties and conductivity can be processed through a chemical route and sintered by SPS [1; 22].

However, due to the complexity of these processing techniques, they are not suitable for large scale production. The planetary milling process has been introduced in their study. Moreover, the details of microstructural development of Si_3N_4 and TiN have not described, especially in the presence of a pulse direct current through the sintering compact during a sintering cycle.

In the present study, we have prepared TiN/Si_3N_4 nanocomposite using SPS from Si_3N_4 and TiN nano powders. The Si_3N_4 and TiN nano powders were applied because they are sensitive to the microstructural changes during the sintering process. The relationship between microstructure and performance, like mechanical properties and electrical conductivity, of these TiN/Si_3N_4 nanocomposites are discussed. Finally, the effect of nano-TiN on the mechanical behavior of Si_3N_4 based nanocomposite has been investigated in sufficient details.

2.1 Experimental Details

2.1.1. Preparation of TiN/Si_3N_4 nanocomposite powder

Commercially available Si_3N_4 nano powder (SM131, Fraunhofer-Institut fur Keramische Technologien and Sinterwerkstoffe, Dresden, Germany) doped with sintering additives of 6 wt% Y_2O_3 and 8 wt% Al_2O_3 was taken as raw material for the matrix phase. It contains 90 wt% β-phase and 10 wt% α-phase, with a manufacturer-determined average particle size of 70nm by the Rietveld method. Nanosized TiN with size of ~30nm (Hefei Kiln Nanometer Technology Development, Hefei, China) was used as secondary phase and it was mixed with Si_3N_4 nano powders. The composition was chosen to yield a TiN content of 5, 10, 15, 20, and 30 wt % in the final product. The specimen designations and corresponding TiN/Si_3N_4 ratios in volume percentage (vol.%) for each composite are shown in Table 1. The mixing powders were ultrasonically dispersed in ethanol for 15 min, and then mixed by planetary milling at a rotation speed of 300rpm for 6 h using a 375 ml nylon bottle with Si_3N_4 balls. The powder mixture was dried in a rotary evaporator, iso-statically cold-pressed into round ingots at a pressure of 200 MPa, crushed and then passed through a #200 sieve for granulation.

TiN/Si_3N_4 content ratio (wt%)	Designation	TiN/Si_3N_4 content ratio (vol%)
5	5TN	3.31
10	10TN	6.75
15	15TN	10.06
20	20TN	14.00
30	30TN	21.81

Table 1. Specimen designation of composites for different TiN content.

2.1.2. Preparation of sintered bodies by SPS

The granulated powders were loaded into a graphite mold with a length of 50 mm and inner and outer diameters of 20 and 50 mm, respectively. A graphite sheet was inserted into the small gap between the punches and mold to improve the temperature uniformity effectively. The graphite mold was also covered with carbon heat insulation to avoid heat dissipation from the external surface of the die. After the chamber was evacuated to a pressure of 10 Pa, the sample was heated to 1600°C under a uniaxial pressure of 30 MPa by SPS (Dr. Sinter 1050, Sumitomo Coal Mining, Kawasaki, Japan). All the SPS measurements were carried out with a heating rate of 200°C/min and holding time of 3 min. A 12 ms-on and 2 ms-off pulse sequence was used. The heating process was controlled using a monochromatic optical pyrometer that was focused on the surface of the graphite mold.

2.1.3. Characterization of sintered bodies

The effective densities of the sintered composites were measured by the Archimedes principle. Phase identification was performed by an X-ray diffractometer (XRD; Model D-MAX/IIB, Rigaku, Tokyo, Japan). Cell dimensions were determined from XRD peak data using UNITCELL with a Si standard. A semiconductor parameter analyzer (HEWLETT PACKARD 4140B, USA) was used to determine the electrical resistivity of the samples. The upper surfaces of the sintered samples were polished down to $1\mu m$. Hardness was measured with a Vickers hardness tester (AKASHI AVK-A, Japan) and by applying a micro-hardness indent at 196N for 15 s. Fracture toughness was measured by the Vickers surface indentation technique [12]. The polished and plasma etched surfaces were used for microstructural characterization by field emission scanning electron microscope (FESEM, XL-40FEG, Philips, The Netherlands). A thin specimen was prepared with a focused ion beam system (FIB, SEIKO, SMI3050, Japan). Transmission electron microscopy (FEGTEM, Tecnai G2 F20, Philips, Eindhoven, Netherlands) was used to characterize the TiN grain of the sintered sample.

2.2. Results and Discussion

2.2.1. Phase Identification of nanocomposite powders by XRD

Fig.2 shows the typical X-ray diffraction patterns of sintered TiN/Si_3N_4 composites with varying TiN content. These composites consist of the β-Si_3N_4 phase as a major phase along with coexistence with secondary TiN phase. The intensity of TiN peaks continues to increase with increasing TiN content. The value of the lattice constant for TiN is 4.25Å, approaching that of pure TiN. On the other hand, the values for a_0 and c_0 for the β-Si_3N_4 phase are 7.61 and 2.91Å, respectively, which are somewhat deviated from those for pure β-Si_3N_4 (+0.01Å). This result suggests that a tiny amount of Si–N may be replaced by Al–O in the particle dissolution and coarsening stages of liquid phase sintering and form the β-SiAlON phase [45].

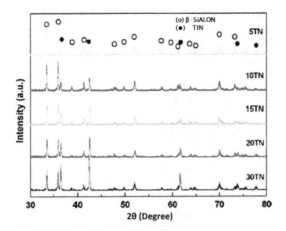

Figure 2. X-ray diffraction patterns of composites with different TiN content.

2.2.2. Densification Behavior

The apparent density of samples containing up to 30 wt% TiN is presented in Fig.3. The apparent density is found to increase with the increase of TiN content. The increase in density is as predicted, because the theoretical density of TiN (5.39 g/cm^3) is substantially greater than that of monolithic Si$_3$N$_4$ (3.19 g/cm^3) (Lide, 2002). No obvious pores are also observed on the polished surfaces of the samples (Fig.4), which suggest the TiN/Si$_3$N$_4$ composites are near full densification.

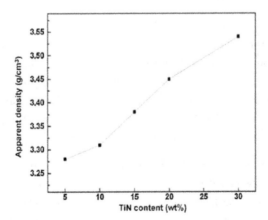

Figure 3. Variation of apparent density of the composites with TiN content.

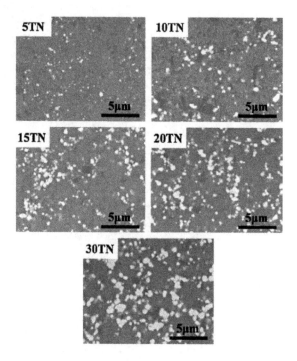

Figure 4. Backscattered SEM images of polished TiN/Si$_3$N$_4$ composites with varying TiN content. The brighter phase is TiN phase and the darker phase is β-SiAlON matrix.

2.2.3. Microstructure Observation of Nanocomposites

The backscattered electron images in SEM of the polished surface of composites with varying TiN content are shown in Fig. 4. As stated earlier, the samples were sintered at 1600°C for 3min with a heating rate of 200°C/min in a vacuum. The lighter and heavier atoms in the backscattered images show up as the gray and white regions corresponding to the β-SiAlON matrix (including glassy phase) and TiN particles. Therefore, the TiN particles are distributed homogeneously in the β-SiAlON matrix. However, most of the TiN appear as submicrosized grains, which are much larger than the size of the starting nano powders.

The typical bright field and dark field images of the TiN grain for the as-sintered composite containing 10 wt% TiN content are shown in Fig. 5(a) and (b), respectively. Fig. 5(c) presents the [0 1 1] selected area diffraction pattern (SAD) for the submicrosized TiN grain, and it shows the existence of a twin structure. The results suggest that grain growth and coalescence of TiN occurs in the composite during the spark plasma sintering process in a short time.

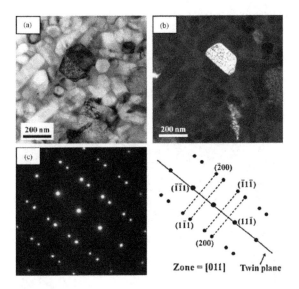

Figure 5. a) Bright field and (b) dark field micrographs, and (c) [0 1 1] selected area diffraction patterns of TiN in spark plasma sintered TiN/Si₃N₄ composite containing 10 wt% TiN.

The typical micrographs of β-SiAlON grains with different TiN content are presented in Fig. 6. In general, the TiN in TiN/Si₃N₄ based composite inhibits grain boundary diffusion and reduces the grain size of the Si₃N₄ matrix [20]. However, for the special case of 10TN among all these composites, the large, elongated grains can be obtained. The conductive phase of TiN might play an important role in the microstructural development of TiN/Si₃N₄ based composites. The electrical resistivity of TiN ($3.34 \times 10^{-7}\Omega$.m) [14] is in the range of metallic materials. Although a tiny current appears as measured by a semiconductor parameter analyzer, it is reasonable that a large current might be induced in the presence of pulsed electrical field during sintering. A leakage current might go through the sintering compact during a heating process, and a similar phenomena is also proposed in ferroelectric ceramics [30; 45], and TiCxOyNz/Si₃N₄ based nanocomposites [31]. Therefore, it is expected that a direct current might hop across conductive TiN grains embedded in the insulating β-SiAlON matrix when applying a pulsecurrent. A temporary high temperature might occur in the specimen, and consequently accelerate the grain coarsening behavior of β-SiAlON during a sintering cycle.

Except for the case of the 10TN composite, most of the β-SiAlON grains for the composites have an equiaxial shape with a grain size of less than 200nm (as shown in Fig. 6), whereas a tiny amount of elongated grains with a grain width of 100nm were observed. For the composite of 5 TN, the percolation concentration is too low (i.e. the interparticle distance of TiN is large) to allow a pulse current to pass through the sintering body [52]. For the samples of 15 TN, 20 TN, and 30 TN, the TiN phase significantly inhibits the grain growth of the β-SiAlON matrix, even though it is possible for a pulse current to pass through the samples during sintering.

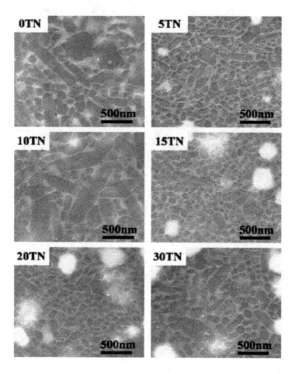

Figure 6. SEM micrographs showing the etching surface of TiN/Si$_3$N$_4$ composites with varying TiN content.

2.2.4. Electrical Properties

The change in electrical resistance for the above composites with varying TiN content is shown in Fig. 7. The electrical resistance substantially decreases from 2.43×10^{10}(5 TN) to 1.93×10^{8}(20 TN) (Ω.m) with the increase in TiN content, whereas it suddenly increases to a value of 4.19×10^{10} (Ω.m) for composite 30 TN. It has been reported that if the fraction of conductive TiN phase in the composite is under the degree of percolation threshold, the insulating property of the composites is maintained as the electrical resistance of non-conductive Si$_3$N$_4$ matrix [52]. This suggests the TiN phase does not form a connective network. As it is evidenced from Fig. 4, the submicrosized TiN grains are nearly isolated from each other. Moreover, the change of electrical resistance possibly depends on the grain size of the conductive phase [11]. Compared with the special cases for 20TN and 30 TN, the larger particle size of TiN for 30TN increases the interparticle distance, leading to a higher magnitude of electrical resistance over composite 20 TN.

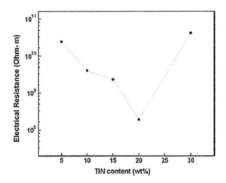

Figure 7. Change of electrical resistance of TiN/Si₃N₄ composites with varying TiN Content.

2.2.5. Mechanical Properties

The mechanical performance of the above nanocomposites has been studied from hardness and toughness measurements. Indentation hardness is a measure of resistance of a sample to permanent plastic deformation due to a constant compression load from a sharp object. This test works on the basic premise of measuring the critical dimensions of an indentation left by a specifically dimensioned and loaded indenter. Vickers hardness is a common indentation hardness scale, which was developed by Smith & Sandland [42]. Vickers test is often easier to use than other hardness tests since the required calculations are independent of the size of the indenter, and the indenter can be used for all materials irrespective of hardness. The influence of the TiN content of TiN/Si₃N₄ based nanocomposites on the Vickers hardness is shown in Fig. 8. The hardness value for these composites decreases with an increasing amount of TiN phase in the β-SiAlON matrix, and a similar trend was also observed by Lee et al., [31].

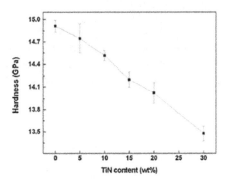

Figure 8. Vickers hardness of TiN/Si₃N₄ composites with various TiN content.

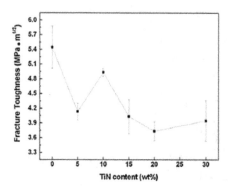

Figure 9. Fracture toughness of TiN/Si$_3$N$_4$ composites measured by Vickers surface indentation technique for various TiN content.

Fracture toughness describes the ability of a material containing a crack to resist fracture, and is one of the most important properties of any material for virtually all design applications. It is also a quantitative way of expressing a material's resistance to brittle fracture when a crack is present. The fracture toughness of the composites containing various compositions of TiN measured by the indentation technique for composites is shown in Fig. 9. The monolithic Si$_3$N$_4$ ceramics have the highest value of 5.4 MPa.m$^{1/2}$. Among the TiN/Si$_3$N$_4$ composites, the toughness reaches a maximum value of 4.9 MPa.m$^{1/2}$ for the composite containing 10 wt% TiN, whereas the other composites have values lower than 4.2 MPa.m$^{1/2}$. An equation derived by Buljan et al., 1988, which expresses the increase in toughness as function of grain size under the assumption that the grain shapes are the same, is given by

$$dK_c = CK_c^0 \left(\frac{dD}{D_0} - 1 \right)$$ (1)

Where C is a coefficient dependent on the mode of fracture; K_c^0 and and D_0 are the initial toughness and grain size; dK_c and dD are the respective changes in toughness and diameter. Hence, the changes in grain size and shape are directly related to toughness. An increase in Si$_3$N$_4$ grain size results in increasing fracture toughness. Although the addition of TiN does not improve the mechanical properties of Si$_3$N$_4$ based composites, the special sintering behavior produced by pulse direct current (grain coarsening effect for Si$_3$N$_4$ based grain) may occur in a SPS process.

3. Conclusions

i.   By utilizing Si$_3$N$_4$ and TiN nano powders as starting materials, a series of near-fully dense TiN/Si$_3$N$_4$ based nanocomposites containing varying TiN contents (5–30 wt %) have been fabricated successfully by a spark plasma sintering technique.

ii.   A grain coalescence of the TiN phase has been demonstrated by TEM. The conductive TiN grains in the insulating Si_3N_4 matrix are observed to be isolated from each other. From the microstructural observations, the composites appear to be insulating materials.

iii.   For the nanocomposite of 5 TN, 15 TN, 20 TN, and 30 TN, the TiN phase inhibits the grain growth of Si_3N_4 based grains during sintering. Hence, the nanosized Si_3N_4 based crystallites are maintained in size as the raw material.

iv.   The spark plasma sintered TiN/Si_3N_4 based composite containing 10 wt% TiN achieves the largest grain size and the highest toughness of 4.9 MPa. $m^{1/2}$ compared to the other composites. A possible pulse current sintering mechanism might occur, which causes a temporary high temperature in the sintering compact, and then accelerates the grain coarsening of Si_3N_4 based grains.

Acknowledgements

Authors are thankful to National Science Council of Taiwan for its financial support under the contract No: NSC 99-2923-E-006-002-MY3 to carry out the present work.

Author details

Jow-Lay Huang* and Pramoda K. Nayak

*Address all correspondence to: JLH888@mail.ncku.edu.tw

National Cheng Kung University, Taiwan

References

[1] Ayas, E., Kara, A., & Kara, F. (2008, July). A novel approach for preparing electrically conductive α/β SiAlON-TiN composites by spark plasma sintering. *Journal of the Ceramic Society of Japan*, 116(1355), 2008, 812-814, 0914-5400.

[2] Buljan, S. T., Baldoni, J. G., & Huckabee, M. L. (1987). Si_3N_4-SiC composite, *American Ceramic Society Bulletin*, 66(2), 347-352, 0002-7812.

[3] Buljan, S. T., Baldoni, J. G., Neil, J., & Zilberstein, G. (1988). Dispersoid-toughened silicon nitride composites,. *ORNL/Sub/8522011/1, GTE*.

[4] Balakrishnan, S., Burnellgray, J. S., & Datta, P. K. (1995). in: S. Hampshire, M. Buggy, B. Meenan, N. Brown (eds.) Preliminary studies of TiN particulate-reinforced

Si(3)N(4) matrix composite (SYALON 501) following exposure in oxiding and oxy-chloridising environments. Key Engineering Matererials, 1013-9826 , 99(1), 279-290.

[5] Borsa, C. E., & Brook, R. J. (1995). Fabrication of Al₂O₃/SiC nanocomposites using a polymeric precursor for SiC. In Ceramic Transactions Ceramic. Processing and Science, The American Ceramic Society, H. Hausner, G. L. Messing and S.-I. Hirano, (Ed.), Westerville, OH , 51, 653-657.

[6] Borsa, C. E., Ferreira, H. S., & Kiminami, R. H. G. A. (1999, May). Liquid Phase Sintering of Al₂O₃/SiC Nanocomposites. *Journal of the European Ceramic Society*, 19(5), 1999, 615-621, 0955-2219.

[7] Blugan, G., Hadad, M., Janczak-Rusch, J., Kuebler, J., & Graule, T. (2005, April). Fractography, mechanical properties, and microstructure of commercial silicon nitride-titanium nitride composites. *Journal of the American Ceramic Society*, 88(4), 2005, 926-933, 1551-2916.

[8] Carroll, L., Sternitzke, M., & Derby, B. (1996, November). Silicon carbide particle size effects in alumina based nanocomposites. *Acta Meterialia,*, 44(11), 1996, 4543-4552, 1359-6454.

[9] Dusza, J., Šajgalík, P., & Steen, M. (1999). Fracture Toughness of Silicon Nitride/Silicon Carbide Nanocomposite at 1350°C. *Journal of the American Ceramic Society*, 82(12), 3613-3615, 1551-2916.

[10] Dusza, J., Kovalčík, J., Hvizdoš, P., Šajgalík, P., Hnatko, M., & Reece, M. (2004). Creep Behavior of a Carbon-Derived Si₃N₄-SiC Nanocomposite. *Journal of the European Ceramic Society*, 24(12), 3307-3315, 0955-2219.

[11] Deepa, K. S., Kumari, Nisha. S., Parameswaran, P., Sebastian, M. T., & James, J. (2009). Effect of conductivity of filler on the percolation threshold of composites. Applied Physics Letters, 0003-6951 , 94(14), 142902.

[12] Evans, A. G., & Charles, E. A. (1976, July). Fracture Toughness Determinations by Indentation. *Journal of the American Ceramic Society*, 59(7-8), 1976, 371-372, 1551-2916.

[13] Greil, P., Petzow, G., & Tanaka, H. (1987). Sintering and hipping of silicon nitride-silicon carbide composite materials, Ceramics International (September 1986) 0272-8842 , 13(1), 19-25.

[14] Gogotsi, Y. G. (1994). Review: particulate silicon nitride based composite. Journal of Materials Science, January (1994). 1573-4803 , 29(10), 2541-2556.

[15] Gao, L., Li, J. G., Kusunose, T., & Niihara, K. (2004, February). Preparation and properties of TiN-Si₃N₄ composites,. *Journal of the European Ceramic Society*, 24(2), 2004, 381-386, 0955-2219.

[16] Galusek, D., Sedláček, J., Švančárek, P., Riedel, R., Satet, R., & Hoffmann, M. (2007). The influence of post-sintering HIP on the microstructure, hardness, and indentation

fracture toughness of polymer-derived Al_2O_3-SiC nanocomposites. *Journal of the European Ceramic Society*, 27(2-3), 1237-1245, 0955-2219.

[17] Guo, Z., Blugan, G., Kirchner, R., Reece, M., Graule, T., & Kuebler, J. (2007, September). Microstructure and electrical properties of Si_3N_4-TiN composites sintered by hot pressing and spark plasma sintering,. *Ceramics International*, 33(7), 2007, 1223-1229, 0272-8842.

[18] Herrmann, M., Balzer, B., Schubert, C., & Hermel, W. (1993). Densification, microstructure and properties of Si_3N_4-Ti(C,N) composites. *Journal of the European Ceramic Society*, 12(4), 287-296, 0955-2219.

[19] Huang, J. L., Chen, S. Y., & Lee, M. T. (1994). Microstructure, chemical aspects and mechanical properties of TiB_2/Si_3N_4 and TiN/Si_3N_4 composites. Journal of Materials Research, 0884-2914 , 9(9), 2349-2354.

[20] Huang, J. L., Lee, M. T., Lu, H. H., & Lii, D. F. (1996). Microstructure, fracture behavior and mechanical properties of TiN/Si3N4 composites, *Materials Chemistry and Physics*, 45(3), 203-210, 0254-0584.

[21] Kawano, S., Takahashi, J., & Shimada, S. (2002). Highly electroconductive TiN/Si_3N_4 composite ceramics fabricated by spark plasma sintering of Si_3N_4 particles with a nano-sized TiN coating. *Journal of Materials Chemistry*, 12(2), 361-365, 1364-5501.

[22] Kawano, S., Takahashi, J., & Shimada, S. (2003, April). Fabrication of TiN/Si_3N_4 ceramics by spark plasma sintering of Si_3N_4 particles coated with nanosized TiN prepared by controlled hydrolysis of $Ti(O-i-C_3H_7)_4$. *Journal of the American Ceramic Society*, 86(4), 2003, 701-705, 1551-2916.

[23] Kašiarová, M., Dusza, J., Hnatko, M., Lenčéš, Z., & Šajgalík, P. (2002). Mechanical Properties of Recently Developed Si3N4 + SiC Nanocomposites. *Key Engineering Materials*, 223, 233-236, 1013-9826.

[24] Kašiarová, M., Rudnayová, E., Kovalčík, J., Dusza, J., Hnatko, M., Šajgalík, P., & Merstallinger, A. (2003). Wear and creep characteristics of a carbon-derived Si_3N_4/SiC micro/nanocomposite. Materialwissenschaft und Werkstofftechnik, April (2003). 1521-4052 , 34(4), 338-342.

[25] Kašiarová, M., Dusza, Ján., Hnatko, M., Šajgalík, P., & Reece, M. J. (2006). Fractographic Montage for a Si_3N_4-SiC Nanocomposite. Journal of the American Ceramic Society May (2006). , 1551-2916 , 89(5), 1752-1755.

[26] Lee, B. T., Yoon, Y. J., & Lee, K. H. (2001, January). Microstructural characterization of electroconductive Si3N4-TiN composites,. *Materials Letters*, 47(1-2), 2001, 71-76, 0016-7577X.

[27] Lide, D. R. (2002). Experimental Data: Evaluation and Quality Control. CRC Handbook of Chemistry and Physics 83rd ed. CRC Press Boca Raton April (2002).

[28] Liu, C. C., & Huang, J. L. (2003). Effect of the electrical discharge machining on strength and reliability of TiN/Si$_3$N$_4$ composites. *Ceramics International*, 29(6), 679-687, 0272-8842.

[29] Liu, X. K., Zhu, D. M., & Hou, W. C. (2006). Microwave permittivity of SiC-Al$_2$O$_3$ composite powder prepared by sol-gel and carbothermal reduction. *Transactions of Nonferrous Metals Society of China*, 16, s494-s497, 1003-6326.

[30] Liu, J., Shen, Z. J., Nygren, M., Kan, Y. M., & Wang, P. L. (2006). SPS processing of bismuth-layer structured ferroelectric ceramics yielding highly textured microstructures. *Journal of the European Ceramic Society*, 26(15), 3233-3239, 0955-2219.

[31] Lee, C. H., Lu, H. H., Wang, C. A., Nayak, P. K., & Huang, J. L. (2011, March). Influence of Conductive Nano-TiC on Microstructural Evolution of Si3N4-Based Nanocomposites in Spark Plasma Sintering. *Journal of the American Ceramic Society*, 94(3), 2011, 959-967, 1551-2916.

[32] Martin, C., Mathieu, P., & Cales, B. (1989). Electrical discharge machinable ceramic composite. *Materials Science and Engineering: A*, 109, 351-356, 0921-5093.

[33] Niihara, K., & Hirai, T. (1986). Super-Fine Microstructure and Toughness of Ceramics. *Bull. Cerum. Soc. Jpn.*, 21(7), 598-604.

[34] Niihara, K., & Nakahira, A. (1988). Strengthening of oxide ceramics by SiC and Si3N4 dispersions. In: Proceedings of the Third International Symposium on Ceramic Materials and Components for Engines, The American Ceramic Society, V. J. Tennery, (Ed.), Westerville, Ohio , 919-926.

[35] Niihara, K., & Nakahira, A. (1991). Strengthening and toughening mechanisms in nanocomposite ceramics. *Ann. Chim. (Paris)*, 16, 479-486, 0151-9107.

[36] Niihara, K. (1991). New design concept for structural ceramics-Ceamic nanocomposites. *The Centennial Memorial Issue of The Ceramic Society of Japan*, 99(10), 974-982, 0914-5400.

[37] Nakahira, A., Sekino, T., Suzuki, Y., & Niihara, K. (1993). High-temperature creep and deformation behavior of Al$_2$O$_3$/SiC nanocomposites. *Ann. Chim. (Paris),*, 18, 403-408, 0151-9107.

[38] Ohji, T., Nakahira, A., Hirano, T., & Niihara, K. (1994). Tensile creep behavior of Alumina/ Silicon carbide nanocomposite. *Journal of the American Ceramic Society*, 77(12), 3259-3262, 1551-2916.

[39] Ohji, T., Jeong-K, Y., Choa-H, Y., & Niihara, K. (1998, June). Strengthening and toughening mechanisms of ceramic nanocomposites. *Journal of the American Ceramic Society*, 81(6), 1998, 1453-1460, 1551-2916.

[40] Raichenko, A. I. (1985, January). Theory of metal powder sintering by an electric-pulse discharge. *Soviet Powder Metallurgy and Metal Ceramics*, 24(1), 1985, 26-30, 0038-5735.

[41] Ragulya, A. V. (2010). Fundamentals of Spark Plasma Sintering. Encyclopedia of Materials: Science and Technology , 978-0-08043-152-9 , 1-5.

[42] Smith, R. L., & Sandland, G. E. (1922). An Accurate Method of Determining the Hardness of Metals, with Particular Reference to Those of a High Degree of Hardness. Proceedings of the Institution of Mechanical Engineers , LCCN 0801 8925, I, 623-641.

[43] Sinha, S. N., & Tiegs, T. N. (1995). Fabrication and properties of Si_3N_4-TiN composite. Ceramic Engineering and Science Proceedings July (1995). , 0196-6219 , 16(4), 489-496.

[44] Šajgalík, P., Dusza, J., & Hoffmann, M. J. (1995). Relationship Between Microstructure, Toughening Mechanisms and Fracture Toughness of Reinforced β-Si_3N_4 Ceramics. Journal of the American Ceramic Society,, 78(10), 2619-2624, 1551-2916.

[45] Shen, Z., & Nygren, M. (2001). Kinetic aspects of superfast consolidation of silicon nitride based ceramics by spark plasma sintering,. Journal of Materials Chemistry, 11(1), 204-207, 0959-9428.

[46] Shen, Z. J., Peng, H., Liu, J., & Nygren, M. (2004). Conversion from nano- to micron-sized structures: Experimental observations. Journal of the European Ceramic Society, 24(12), 3447-3452, 0955-2219.

[47] Šajgalík, P., Hnatko, M., Lojanová, Š., Lenčéš, Z., Pálková, H., & Dusza, J. (2006). Microstructure, hardness, and fracture toughness evolution of hot-pressed SiC/Si_2N_4 nano/micro composite after high-temperature treatment. International Journal of Materials Research , 1862-5282 , 97(6), 772-777.

[48] Vollath, D., Szabó, D. V., & HauBelt, J. (1997). Synthesis and Properties of Ceramic Nanoparticles and Nanocomposites. Journal of the European Ceramic Society,, 17(11), 1317-1324, 0955-2219.

[49] Vollath, D., & Szabó, D. V. (2006). Microwave Plasma Synthesis of Ceramic Powders. In: Advances in Microwave and Radio Frequency Processing Part-IX, Springer , 619-626.

[50] Xu, Y., Nakahira, A., & Niihara, K. (1994). Characteristics of Al_2O_3-SiC nanocomposite prepared by sol-gel processing. Journal of the Ceramic Society of Japan, 102(3), 312-315, 0914-5400.

[51] Yoshimura, M., Komura, O., & Yamakawa, A. (2001, May). Microstructure and tribological properties of nano-sized Si 3N4. Scripta Materialia, 44(8-9), 2001, 1517-1521, 1359-6462.

[52] Zivkovic, L., Nikolic, Z., Boskovic, S., & Miljkovic, M. (2004, June). Microstructural characterization and computer simulation of conductivity in Si3N4-TiN composites,. Journal of Alloys and Compounds, 373(1-2), 2004, 231-236, 0925-8388.

Chapter 7

Study of Multifunctional Nanocomposites Formed by Cobalt Ferrite Dispersed in a Silica Matrix Prepared by Sol-Gel Process

Nelcy Della Santina Mohallem, Juliana Batista Silva,
Gabriel L. Tacchi Nascimento and
Victor L. Guimarães

Additional information is available at the end of the chapter

1. Introduction

Surface science has a long history, involving the development of colloids, particulate material, thin films and porous materials. These materials have been known and used for centuries, without a profound knowledge of their real physical–chemistry characteristics. The detailed study of their properties was only possible with the emergence of more sophisticated spectroscopic techniques, and high-resolution eletron microscopes [1].

With the development of nanoscience and nanotechnology, new materials began to be studied like nanoparticles, porous materials that are formed by a network of nanoparticles, and nanocomposites. There are infinite possibilities of production of nanocomposites and one of them is the formation of nanoparticles inside porous matrices that can have their texture and morphology tailored by thermal treatment or templates [2].

For many applications, the textural properties, such as porosity and specific surface area, are as important as the chemical composition. Thus, the growing demand for porous products in the industry of nanotechnology, especially for magnetic nanocomposites, has led to the increase in the studies related to these properties [1, 2].

Magnetic materials have been used by man for centuries, since ancient people discovered the natural magnets called lodestones. The term "magnet" is used for magnetic materials that produce their own magnetic field. Other magnetic materials have magnetic proprieties only in response to an applied magnetic field. There are several types of magnetic materials that have been used in diverse devices and systems for industrial products. Some traditional applications of these materials are in cores for motors, generators and transformers, microwave devices, magnetic media used in computers, recording devices, and magnetic cards, among others [3].

There are various metallic elements (Fe, Ni, etc) that have magnetic properties due to their crystalline atomic structure whose spins align spontaneously. Some alloys formed by metallic elements and others including the earth rare elements also have excellent magnetic properties (alnico, samarium-cobalt and neodymium-iron-boron magnets). Finally, the ferrites are a known class of magnetic materials formed by metallic oxides.

With the advancement of the material sciences and the emergence of the nanoscience and nanothecnology, new kinds of magnetic materials have been developed and studied in the last years, such as the magnetic nanoparticles, ferrofluids and magnetic nanocomposites. With these materials, new applications could be tested in areas such like electronic, catalysis and biomedicine, among others [4].

1.1. Ferrites

Ferrites are chemical compounds obtained as powder or ceramic body with ferrimagnetic properties formed by iron oxides as their main component, Fe_2O_3 and FeO, which can be partly changed by others transition metals oxide. The ferrites can be classified according their crystalline structure: hexagonal ($MeFe_{12}O_{19}$), garnet ($Me_3Fe_5O_{12}$) and spinel ($MeFe_2O_4$), where Me represents one or more bivalent transition metals (Mn, Fe, Co, Ni, Cu, and Zn). The ferrites are classified as "soft" or "hard" magnets, according to their magnetic properties, which refers to their low or high magnetic coercivity, respectively. Hard magnets are not easily demagnetized (curve a), due to their high coercivity and soft magnets are easily magnetized and demagnetized (curve b) with application of a magnetic field, due to their low coercivity. The characteristic magnetic hysteresis curves of these type of magnets are shown in Figure 1 [3,5,6].

The intermediate magnets, generally used in magnetic media, must have coercivity sufficiently high for withholding the information, but sufficiently low to allow for the information to be deleted (curve c) [5,6].

These magnetic ceramics [6] are important in the production of electronic components, since they reduce energy losses caused by induced currents and they act as electric insulators. They can be used in simple function devices such as small permanent magnets, until as sophisticated devices for the electro-electronic industry.

Recently, these materials have been discovered as good catalysts [7,8,9] and biomaterials [10,11].

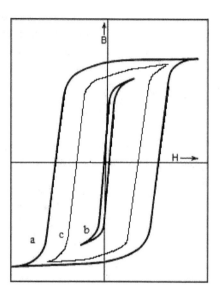

Figure 1. Magnetic hysteresis curves of (a) hard, (b) soft and (c) intermediate magnets.

1.1.1. Cobalt ferrites

Cobalt ferrite [3, 6], an intermediate magnet, is an important multifunctional magnetic material not only for its magnetic properties but also for its biomedical and catalytic applications, which depend on their textural and morphological characteristics. Cobalt ferrite, that has great physical and chemical stability, has been used in the production of permanent magnets, magnetic recording such as audio and videotape and high-density digital recording disks, magnetic fluids and catalysts. This ferrite has spinel inverse structure and exhibits a large coercivity, differently from the rest of the spinel ferrites. The magnetization of the $CoFe_2O_4$ crystal has anisotropic character because depends on its orientation. The strong magnetic flux promoted by the superexchange interaction is directed along of the magnetization direction, and generally may be coinciding with the crystallographic axes. The magneto-crystalline anisotropy is related with the spin-orbit coupling. In polycrystalline materials, the magnetization measured corresponds to a mean value.

Recently these metal-oxide nanoparticles have been the subject of much interest because of their unusual optical, electronic and magnetic properties, which often differ from the bulk. These nanoparticles should have single domain, of pure phase, having high coercivity and intermediary magnetization [12].

The properties of the cobalt ferrite are changed according to the form of obtainment of the material, as bulk, particles or nanoparticles. The nanocrystalline particles have a high surface/volume ratio, and thus, they present different properties from those of bulk materials

Various authors [3,4,6,12,13] described the saturation magnetization and coercivity measured at room temperatures as a function of crystallite size and these values change from 30 to 80 emu g^{-1} for saturation magnetization and of 0.5 to 5.4 kOe for coercivity for crystallite size varying from 4 to 50 nm.

The effect of thermal vibrations is largest in very small particles, especially in materials with low anisotropy. The magnetic moments assume random orientations, at room temperature, for nanoparticles with size below the limit of 4-10 nm, resulting in superparamagnetic behavior [14-16].

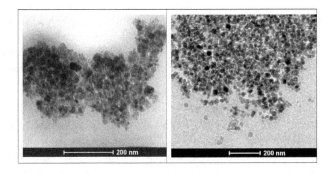

Figure 2. Cobalt ferrite nanoparticles obtained by (a) coprecipitation and (b) hydrothermal processes.

Superparamagnetic materials magnetize and demagnetize more easily than the other ones due their dimension being equivalent to a magnetic domain. The magnetic domain of very small particles is different from that observed in bulk structures. There is a critical diameter below which the formation of a domain wall results in an increase of the total energy. The mono-domain size for $CoFe_2O_4$ nanoparticles has been estimated between 10 and 70 nm [17]. Crystallites with diameter smaller than 10 nm have superparamagnetic behavior, while with diameters larger than 70 nm (critical particle size/Dc) show multi-domain microstructure, with the consequent decrease in coercivity [17]. The existence of multiple domains separated by walls governs the magnetic behavior. The magnetization and demagnetization processes driven by an external field are characterized by the nonexistence of the hysteresis, characteristic of superparamagnetic materials.

Because of these interesting characteristics, nanocrystalline ferrites have been extensively studied with emphasis on the particle size variation and the influence of this variation in the mechanical, biomedical, magnetic and catalytic properties. In order to achieve desired properties, it is necessary to obtain high-density powders with a small and uniform grain size, and controlled stoichiometry. Hence, there is the need to develop fabrication processes relatively simple that induce the formation of controlled particle size materials. Some nanoparticles can be achieved more easily by using chemical methods [3,4,8,12,18,19], such as coprecipitation, hydrothermal synthesis and sol-gel process, among others, but generally the nanoparticles tend to agglomerate due to their high reactivity. Figure 2 shows some cobalt ferrite nanoparticles obtained by coprecipitation and hydrothermal processes.

Due to these problems with high reactivity, agglomeration and aggregation of the nanoparticles, and the possibility of the development of new materials with peculiar properties, it has been synthesized nanocomposite materials formed by metal or metallic oxide nanoparticles dispersed in ceramic or vitreous matrices, avoiding the agglomeration and improving the dispersion and the distribution of the nanoparticles inside the system [20-34].

1.2. Nanocomposites

A composite is considered as a multiphase material with significant proportion of the properties of the constituent phases, whose final product has its property improved. There is the possibility of combining various types of materials in a single composite, in order to optimize their properties according to the desired application. When one of the phases has nanometric dimension, the system is called nanocomposite. [25,34].

Nanocomposite materials formed by metal or metallic oxide nanoparticles dispersed in ceramic or vitreous matrices have important applications due to the possibility of developing more reactive materials with new properties. The interest in the preparation of magnetic nanocomposites has increased in the last years due to the properties presented by these materials, which depends on the particle size, concentration and distribution of the particles in the matrix. Nanosystems such as Fe/SiO_2, Ni/SiO_2, Fe_3O_4/SiO_2, $CoFe_2O_4/SiO_2$, $NiFe_2O_4/SiO_2$ have been intensively studied in the last years, revealing different behavior from those of bulk magnetic systems and serving as models for the study of small particles [25-34].

The texture of the matrix and the interaction between the magnetic nanoparticles and the host matrix can be used to control the magnetic properties and the stability of these materials.

The magnetic nanoparticles dispersed in a inert matrix act as isolated nanomagnets, eliminating energetic losses, and producing the coupling between neighboring nanoparticles, which improve their magnetic properties. These nanocomposites can have high chemical and structural stability, high catalytic activity and high mechanical resistence.

The crystallite size control inside the matrix is justified by the existence of an average diameter range of single domain crystallites, between 10 nm < d < 80 nm, depending on the desired optimal magnetic properties [36,37]. When the ferrite concentration is low (< 30%), the crystallites are isolated, having single domains and showing superparamagnetism. Concentrations above 50% of ferrite provoke the agglomeration of the crystallites, which results in multi-domains [20]. Another important characteristic of nanocomposites in general is the texture of the matrix, as pore distribution, pore size and specific surface area, which has large influence in their final characteristics, as the transport and interaction of fluids within their connected network formed by micro, meso and macropores [38-41]. Important materials to be used as porous matrices are xerogels and aerogels, material obtained by sol-gel process.

In the last years, the sol-gel process has been used to produce magnetic nanocomposites by incorporation of ultra-fine magnetic nanoparticles with a high surface/volume ratio in different matrices. The nanocomposites formed have different properties from the magnetic bulk.

1.3. Sol-gel process

The sol-gel chemistry is based on mechanisms of hydrolysis and condensation of precursors containing metal (s) of interest, called "sol", resulting in an M-O-M oxide network that form a wet gel. There are two types of precursor: an aqueous solution of an inorganic salt or a metal alcoxide compound. The gel may be formed by polymerisation (gel polymer) or aggregation of colloidal particles subject to the physic-chemical conditions of the medium (colloidal gel). In either case, a three-dimensional solid network of the gel retains a liquid phase in its pores [42-49].

In practice, the network structure and the morphology of the final product depend on the relative contributions of the reactions of hydrolysis and condensation. These contributions may be controlled by varying the experimental conditions: the type of metal, type of organic binder, the molecular structure of the precursor, water/alkoxide ratio, type of catalyst and solvent, temperature and concentration of the alkoxide.

After the gelification, the wet gel is subjected to aging, to occur the polymerization processes, syneresis, and neck formation between the particles, leading to increase in connectivity and strength of the gel structure. The gel obtained is formed by a solid structure impregnated with the solvent. After the aging, various drying processes can be used to convert the wet gel in a porous material, denominated xerogel or aerogel.

The sol-gel process allows the preparation of materials in various forms such as powders, thin films, and monoliths, with desirable properties such as hardness, chemical durability, thermal and mechanical resistance and with different textures. The final product (xerogel or aerogel) can be tailored by different temperatures of thermal treatment leading to materials with different specific surface areas and porosities.

1.4. Xerogels and aerogels

The drying is one of the more important steps in sol-gel process because it is possible to obtain different materials by changing the drying routes. During the drying, the solvent adsorbed inside the porous gel is removed. During this process the gel network can collapse.

There are several types of drying processes; among them we can mention the controlled drying and the supercritical drying. In the controlled process, the solvent is evaporated slowly at room temperature and pressure, generating a contraction on the material, provoking the decreasing in the pore size due to the surface tension. The dry gels obtained by this process are called xerogels and have high porosity and specific surface area [27,50,51].

In the supercritical drying, the wet gels are put in a reactor at high temperature and pressure, above the critical point of the system, where there is no discontinuity between the liquid and gaseous phase, avoiding capillary forces. The dry gels obtained are called aerogels and have higher porosity than the xerogel. [27,52,53].

In this work we studied the characteristics of nanocomposites formed by cobalt ferrites dispersed in silica matrix (xerogel and aerogel) obtained at different thermal treatment temperatures. Techniques such as X-ray diffraction (XDR), spectroscopy in the infra-red region, force atomic microscopy, transmission electron microscopy, scanning electronic microscopy equipped with energy dispersive X-ray (EDS) and wavelength dispersive (WDS) probes and gas adsorption were used to study the morphological and structural changes of the materials as a function of the thermal treatment temperature. The results were used to evaluate the mechanism of formation of the nanocomposites and relate their characteristics with magnetic and catalytic properties.

2. Experimental

2.1. Silica Matrix and Cobalt Ferrite Nanocomposite

The inert matrices formed by porous pure silica were obtained by mixing tetraethylorthosilicate, ethyl alcohol, water (1/3/10) and nitric acid, used as a catalyst. The nanocomposite precursor solution was obtained by mixing cobalt and iron nitrates, $(Co(NO_3)_2.6H_2O$ and $Fe(NO_3)_2.9H_2O)$ with the matrix precursor, to form nanocomposites with 30 wt% of ferrite. The solutions were stirred for 1 h for homogenisation and left to rest for gelation, which takes place due to the hydrolysis and polycondensation of the metallic alkoxides. The wet gels were aged at 60 °C for 24 h and dried at 110 °C for 12 h, leading to the formation of xerogels. Aerogels were formed by supercritical drying of the wet gels in an autoclave under 180 atm of N_2 and raising temperature up to 300 °C at 5 °C min^{-1}, temperature and pressure adequate to exceed the critical point of the mixture ethyl alcohol/water. The system was kept in this condition for 2 h. All xerogel and aerogel were heated between 300 and 1,100 °C for 2 h.

2.2. Caracterization Techniques

The structural evolution of the samples were analyzed in an X-ray diffractometer (Rigaku, Geigerflex 3034) with CuKa radiation, 40 kV and 30 mA, time constant of 0.5 s and crystal graphite monochromator. Crystallite sizes were determined by Scherrer equation (D = 0.9k/b cos h, where D is the crystallite diameter, k is the radiation wavelength and h the incidence angle). The value of b was determined from the experimental integral peak width using silicon as a standard. The values were corrected for instrumental broadening.

Spectra in the infrared region were obtained in an ABB Bomem equipment, model MB 102.

The composite compositions were evaluated by an electron microprobe (Jeol JXA, model 8900RL) with energy dispersive (EDS) and wavelength dispersive (WDS) spectrometers.

The morphology was obtained by scanning electron microscopy (high resolution SEM - Quanta 200 - FEG - FEI), by transmission electron microscopy (high resolution TEM - FEI) and by atomic force microscopy (Dimension 3000, Digital Instruments Nanoscope III - LNLS).

Sample textural characteristics were determined by N_2 gas adsorption (Quantachrome, model Nova 1200) at liquid nitrogen temperature. The samples dried at 110 °C were outgassed at 100 °C for 3 h. The others ones were outgassed at 200 °C for 3 h before each experiment. Specific surface areas and total pore volumes were obtained by the Brunauer-Emmett-Teller (BET) equation and the Barrett-Joyner-Halenda (BJH) method. These measurements were used to evaluate the total porosity, by the equation $P = 1 - \varrho_{ap}/\varrho_{th}$, where ϱ_{ap} is the apparent density and ϱ_{th} is the theoretical density. True densities were obtained in a helium picnometer (Quantachrome) and apparent densities were obtained by mercury picnometry.

The magnetic measurements were made in a Lake Shore vibrating sample magnetometry (VSM) at 300 K with a maximum applied magnetic field of 1 Tesla.

The nanocomposites were tested as catalysts in the oxidation of chlorobenzene in air. The catalytic reactions were carried out in a fixed bed reactor with 25 mg of catalyst. Chlorobenzene at 0.1% was introduced in the air stream (30 mL min-1) by a saturator at 0°C. The reaction products were analyzed by gas chromatography (Shimadzu/GC 17A) with a flame ionization detector (FID) and an Alltech Econo-Cap SE capillary column (30 mm 90. mm 9 0.25 lm).

3. Results and Discussion

The xerogels and aerogels silica matrices and nanocomposites were obtained in the monolithic form, without cracks.

3.1. Structural characterization

SiO_2 xerogel and aerogel treated up to 900°C exhibit amorphous behavior. A narrowing of the XRD peak accompanied by an increase in its intensity with increasing in the temperature of the preparation indicate an increase in the structural organization of the samples (Figure 3a and 3b). Characteristic reflections of crystobalite and tridimite appear at 1100°C for both the samples. The intensity of the xerogel peaks are larger than the aerogel ones.

The xerogel nanocomposites exhibit amorphous behavior up to 300 °C (according the X-ray diffractometer resolution). $CoFe_2O_4$ crystalline particles with cubic spinel structure are detected by XRD inside the amorphous silica matrix above this temperature (Figure 4a). During the formation of the $CoFe_2O_4$ nanocrystals, no traces of intermediate products are found even at temperatures as high as 1100°C, indicating that the ferrite particles were formed without binding to the matrix. The magnetic nanoparticles also avoided the formation of either crystobalite or tridimite phases.

The aerogel nanocomposites exhibit amorphous behavior up to 700 °C (Figure 4b). The $CoFe_2O_4$ phase is detected by XRD only above this temperature.

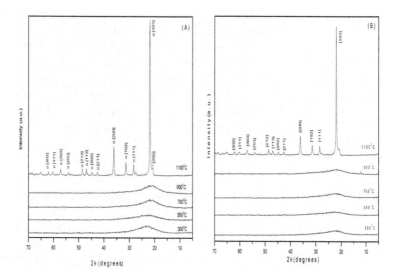

Figure 3. X-ray diffraction patterns of SiO₂ (a) xerogel and (b) aerogel, thermally treated in air for 2 hours at various temperatures. Crystobalite (o) and tridimite (*)

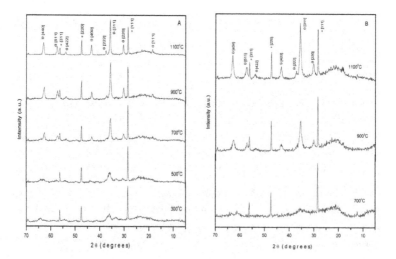

Figure 4. X-ray diffraction patterns of CoFe₂O₄/SiO₂ (a) xerogel and (b) aerogel, thermally treated in air for 2 hours at various temperatures. (Si) silicon, (Co) CoFe₂O₄.

Figure 5a shows the IR spectra of the xerogel samples obtained after heat-treating of the dried gel at various temperatures for 2 hours. The IR spectrum of the sample dried at 300°C has absorptions characteristic of the silica network at 1080, 810, and 460 cm^{-1}. The 1086 cm^{-1} band with the shoulder at 1160 cm^{-1} is due to the asymmetric stretching bonds Si-O-Si of the SiO$_4$ tetrahedron associated with the motion of oxygen in the Si-O-Si anti-symmetrical stretching. The 810 cm^{-1} band is associated with the Si-O-Si symmetric stretch and the band at 461 cm^{-1} with either Si-O-Si or O-Si-O bending. The weak band in the 950cm^{-1} is due to stretching of the Si-OH. With increasing in temperature this band disappears due to the condensation reactions which change the Si-OH groups on Si-O-Si. The sample heated at 900°C shows a decrease in intensity of the bands characteristic of the silica matrix, suggesting the rearrangement process of silica network, according XRD results. The aerogels have similar spectra.

Figure 5b shows the IR spectra of the xerogel nanocomposites. The IR spectrum of the sample dried at 300°C also has absorptions characteristic of the silica network and at 968 cm^{-1} we observed the band composed of the contributions from Si-O-H and Si-O-Fe vibrations. The band at 584 cm^{-1} is related to Fe-O stretching. The 968 cm^{-1} band disappears with the increase in temperature, showing that the weak bond between Si and Fe is broken [54]. We can observe a slight shift of the 584 cm^{-1} band to the left. Co-O stretching vibration characteristic band also appear at 461 cm^{-1}. The weak band at 675 cm^{-1} can be due to the cobalt ion in tetrahedral centers in the matrix pores.

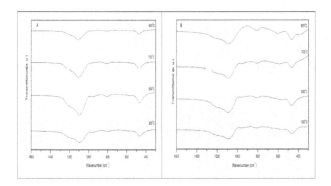

Figure 5. a) Infrared spectra of SiO$_2$ xerogel and (b) infrared spectra of CoFe$_2$O$_4$/SiO$_2$ xerogel treated at different temperatures.

IR spectra show that there was no formation of by-product, confirming X-ray diffraction results. All results suggest that the iron ions had interaction with the silica matrix when the nanocomposite was heated at low temperatures. These interactions disappeared with increasing in heating temperature, showing that they are weak bonds. These results suggest that the cobalt ions have been diffused by the porous matrix and have been bonded to iron to form the ferrite within the pores, without any binding or interaction with the silica network.

3.2. Textural Characteristics

3.2.1. Silica xerogels and nanocomposite xerogels

The xerogel is a typical porous material formed by a silica network with micro, meso and macro pores interconnected for all the bulk. Micro pores are pores smaller than 2 nm in diameter, meso pores are the pores with diameters between 2 and 20 nm, and macro pores are larger than 20 nm.

Monolithic porous matrices (Figure 6), without defects after drying, changed in size after thermal treatments at high temperatures. The shape of the samples were defined by the template. The silica xerogels are optically transparent in all temperatures of preparation.

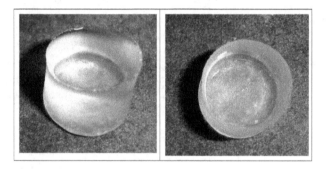

Figure 6. Silica xerogel.

AFM and TEM images (Figure 7) show the microstructure of a typical xerogel.

Figure 7. a) atomic force microscopy and (b) transmission electron microscopy images of a typical silica xerogel prepared at 500 °C.

The textural characteristics of the silica matrix xerogel changed substantially with thermal treatment. The specific surface area (Figure 8) decreased gradually for samples prepared above 300°C, and between 500 and 700°C decreased rapidly due to the densification process of the material. The silica xerogel porosity (Figure 9) remained constant for samples prepared up to 700 °C and decreased sharply until 900 °C, due to the collapse of the pores. The sample heated at 1100°C became a material with a continuous silica network without pores.

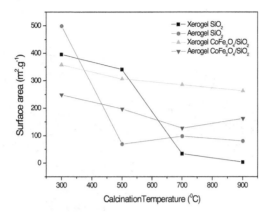

Figure 8. Variation of surface area as a function of temperature for composite CoFe$_2$O$_4$ in SiO$_2$ matrix and SiO$_2$. Error: 5%

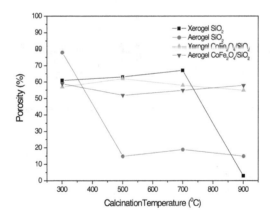

Figure 9. Variation of the porosity of silica matrices and cobalt ferrite nanocomposites as a function of temperature. Error: 5%

The formation of the ferrite nanoparticles inside the pores of the xerogel matrix reinforced the silica structure, keeping stable the pore network at high temperatures. In this case, the specific surface area decreases about 14% (Figure 8) and the total porosity remain almost constant between 300°C and 900°C (Figure 9). The shrinkage of the material structure occured only in samples heated above 900 °C.

Figure 10 shows the microstructure of the ferrite nanocomposite. With the increase in temperature, the ferrite grows inside the silica matrix.

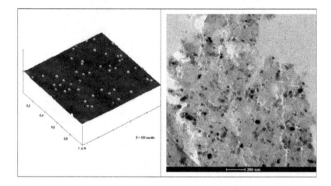

Figure 10. Atomic force microscopy and transmission electron microscopy images of cobalt ferrite nanocomposite prepared at 900 °C.

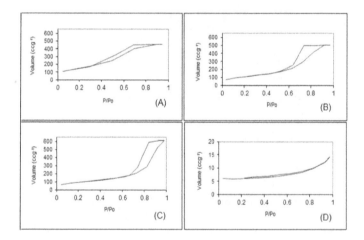

Figure 11. Adsorption-desorption isotherms of silica xerogels heated at (a) 300°C, (b) 500°C, (c) 700°C and (d) 900°C.

Figure 11 shows the adsorption-desorption isotherms of the SiO_2 xerogel at different thermal treatment temperatures. The sample prepared at 300°C adsorbed about 450 cm³.g⁻¹ of N_2. The xerogel adsoved more gas with increasing in the heating temperature due to the liberation of the organic compounds of their pores. The xerogel prepared at 700°C adsorbed about 600 cm³.g⁻¹. The isotherm of the sample heated at 300 °C presented characteristic intermediary of meso and microporous materials. The isotherms of the samples treated at 500°C and

700°C presented mesoporosity characteristics (isotherm type IV according the BDDT classifi-
cation [39,41]). The samples heated at 900°C presented isotherm type III, without hysterese,
characteristic of non-porous material.

With the formation of ferrite nanoparticles inside the xerogel the material prepared at all
temperatures became mesoporous corroborating the reinforcement in the xerogel microstructure.

All nanocomposite xerogel isotherms shown in Figure 12 are characteristic of mesoporous
materials. The nanocomposite without thermal treatment adsorbed about 200 cm³/g of N₂,

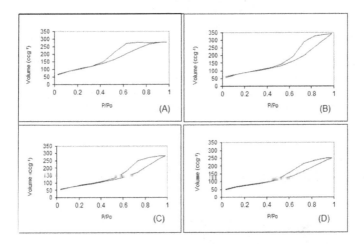

Figure 12. Adsorption/desorption isotherms of CoFe₂O₄ / SiO₂ xerogel, heated at (a) 300°C, (b) 500°C, (c) 700°C and
(d) 900°C.

while nanocomposites prepared at higher temperatures adsorbed about 300 cm³/g of N₂, due
to the elimination of solvents. All isotherms (type IV [39,41]) presented characteristics of
mesoporous materials. The hysteresis curves are type H2, characteristics of pores with indef-
inite form and size, according to the AFM and TEM images (Figure 10).

3.2.2. Silica aerogels and nanocomposite aerogels

The aerogel is an extremely porous material also formed by a silica network with micro,
meso and macro pores interconnected for all the bulk. This material is much more porous
than the xerogel, mainly as prepared. Monolithic aerogel matrices (Figure 13), without defects after drying, changed sharply in size after thermal treatments at high temperatures.
The silica aerogels are slightly opaque due to the macroporosity, whose pores with the same
size than the light wavelength interfere with the optical transparence.

Figure 13. Silica aerogel.

Figure 14a shows a like-smoke structure of the aerogel. It is very difficult to obtain SEM and AFM images of this kind of material, due to their high porosity. Almost any preparation can destroy the network formed by the interconnection of the nanoparticles. Figure 14b shows the porous network structure of the aerogel, evidencing the necks formed by silica nanoparticles that led to the formation of micro, meso and macropores.

The as-prepared silica aerogels presented higher specific surface area and porosity than the xerogels obtained from the same precursor (Figure 8 and 9). Nevertheless, contrary to what happens with the xerogel, the porous network collapsed at low temperature of preparation

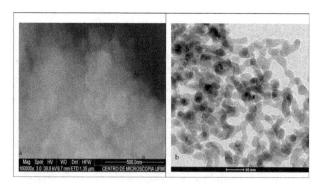

Figure 14. Scanning (a) and transmission (b) electron microscopies images.

of about 500 ^{0}C. The specific surface area and porosity changed from 500 m^{2}/g and 80 % for aerogels prepared at 300 ^{0}C to 50 m^{2}/g and 15 % for aerogels prepared at 500 ^{0}C. These values have been kept constant until temperatures of about 900 ^{0}C, when occurred the total collapse of the porous.The formation of the ferrite nanoparticles inside the pores of the aerogel matrix also reinforced their silica structure as happened with the xerogel matrices, but the values of specific surface area were lower than the xerogel ones, as seen in Figure 8. In this case, the specific surface area Decreased of 250 to 150 m²/g. The aerogel porosity values increased in comparison to the silica aerogel matrix and remained almost constant between 300°C and 900°C, with similar values to the xerogel ones. The shrinkage of the material network also occured above 900 °C.

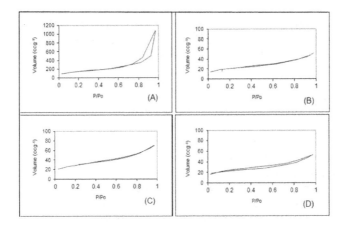

Figure 15. Adsorption-desorption isotherms of silica aerogels heated at (a) 300°C, (b) 500°C, (c) 700°C and (d) 900°C.

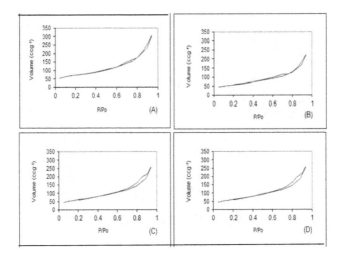

Figure 16. Adsorption/desorption isotherms of CoFe₂O₄ /SiO₂ aerogel obtained for sol-gel method and heated at (a) 300°C, (b) 500°C, (c) 700°C and (d) 900°C.

Figure 15 shows the silica aerogel isotherms. The isotherm of the sample prepared at 300ºC presented characteristics of mesoporous material (39,41), with large N_2 adsorption, of about 1200 $cm^3.g^{-1}$. This material is very fragile and when submitted at high heating temperatures its structure was annihilated, adsorbing only 60$cm^3.g^{-1}$ when prepared between 500 and 900 ºC. Samples prepared between these temperatures did not present hysteresis and the adsorption-desorption curves are characteristics of macroporous materials [39,41].

All nanocomposite aerogel isotherms shown in Figure 16 are characteristic of macroporous materials. The as-prepared aerogel nanocomposite adsorbed ~300 $cm^3.g^{-1}$ of N_2, value much lower than that presented by the silica aerogel. The nanocomposites prepared at higher temperatures adsorbed about 250 $cm^3.g^{-1}$ of N_2, values larger than the silica aerogels prepared in similar temperatures.

3.2. Mechanism of formation

Backscattered electron images of the cobalt ferrite nanocomposites prepared between 300 and 700 ºC showed defined white regions distributed throughout the sample, whose EDS analyses detected mostly the presence of cobalt clusters. In the gray region, Si, Fe, O and traces of Co were detected (Figure 17). The clusters disappeared with the increase in the preparation Temperature, suggesting that the cobalt ions diffused into the composite, binding to iron to form the ferrite. At temperatures above 900ºC, EDS analyses detected a homogeneous distribution of Co and Fe in the composite. These results corroborate XRD and IR results.

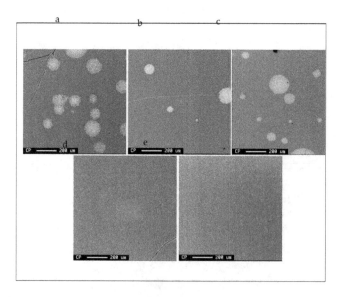

Figure 17. SEM images (backscattered electrons) of xerogels prepared at (a) 300ºC, (b) 500ºC, (c) 700 ºC, (d) 900 ºC and (e) 1100 ºC.

Figure 18 shows the WDS mapping, used to confirm the mechanism of formation of nano-particles inside the porous silica matrix. The different concentration of each metallic ion is shown by the color evolution. By this mapping, it is possible to observe with more acuity the diffusion process of the ions as a function of the temperature of preparation.

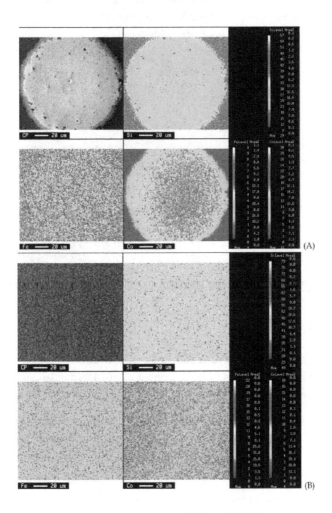

Figure 18. WDS mapping of the nanocomposite prepared at (a)500°C and (b)1100°C

Figure 19 shows the proposed model of the formation mechanism of the cobalt ferrite nano-composites. The precursor solution is prepared by the mixing of Si alkoxide, alcohol, water and Fe and Co nitrates, and some catalysts (Figure 19 a). The wet gels are formed by hydrolysis and polycondensation of the sol constituents, maintaining the same ions distribution.

After drying, the elimination of water and organic residues occurs, and the xerogel (or aerogel) is formed by a silica network with iron ions distributed by the network and weakly bonded to the silicon (Figure 19 b). The Co ions are agglomerated in definite regions forming the clusters. With increasing temperature of preparation, the cobalt ions diffuse by the silica network, forming a chemical bond with the iron, which has its weak bond with the silicon broken (Figure 19 c). The pores diminish in amount and in size, and the magnetic nanoparticles grow inside these pores with the increase in temperature, leading to the encapsulation of the nanoparticles by the silica matrix (Figure d and e).

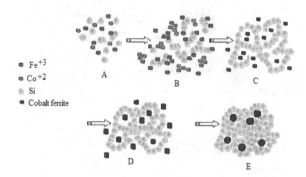

Figure 19. A proposed model of the formation mechanism of cobalt ferrite nanocomposites.

3.3. Properties and applications

Magnetic xerogels and aerogels can be considered as multifunctional materials due to the possibility of use their properties in multiple applications. The desired application can be obtained by tailoring the characteristics of the material.

Multifunctional materials are composites or systems capable of performing multiple functions simultaneously, depending of the involved phases, their structural, morphological and textural characteristics, improving system performances and reducing the redundancy between the composite components and their individual functions.

For example, porous xerogel and aerogel nanocomposites have interesting catalytic properties when tested in the oxidation of chlorobenzene in air. Figure 20 shows the performance of the xerogel nanocomposites prepared at various temperatures, compared to SiO_2. It is clear that the more porous nanocomposite had the best performance. Figure 21 shows the best performance of the xerogel and the aerogel prepared at 500 °C compared with ferrite powders obtained by coprecipitation.

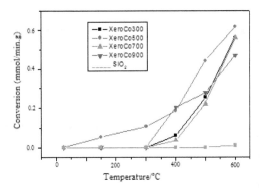

Figure 20. Catalytic oxidation of chlorobenzene in the presence of CoFe$_2$O$_4$/SiO$_2$ xerogels thermally treated at various temperatures.

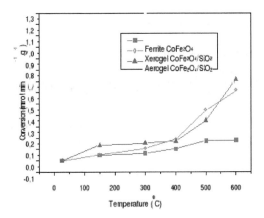

Figure 21. Catalytic oxidation of chlorobenzene in the presence of CoFe$_2$O$_4$/SiO$_2$ xerogels, aerogel and ferrite.

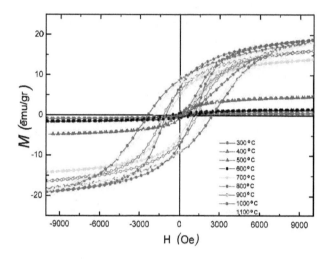

Figure 22. Hysteresis curves of the xerogel nanocomposites prepared at several temperatures.

The magnetic properties also are tailored by temperature. Figure 22 shows the hysteresis curve of the cobalt ferrite nanocomposites prepared at several temperatures, whos exerogels showed since superparamagnetic characteristics when prepared at low temperatures and intermediate magnetism at high temperatures. Due to these properties, these nanocomposites can be used as electronic devices or in biomedicine in cancer treatment by hyperthermia and drug release controlled by magnetic field.

4. Conclusion

Nanocomposites formed by cobalt ferrite nanoparticles dispersed in porous silica matrix ($CoFe_2O_4/SiO_2$) were prepared by the sol-gel process. The presence of the magnetic nanoparticles inside of the inert porous matrices of xerogels and aerogels reinforced their structure, avoiding large changes in specific surface area, porosity and in the microstructure of the matrix after preparation temperature, which varied between 300 and 900 °C. These characteristics influence the properties of the nanocomposites, such as their chemical reactivity, catalytic activity and magnetization. Due to the possibility of tailoring the textural and morphological characteristics of these types of multifunctional nanocomposites, they are promising candidates for many technological applications in electronic, catalysis and biomedical areas.

Acknowledgements

This work was supported by CNPq and Fapemig (Brazilian agencies). The authors acknowledge the use of the infrastructure of the Centre of Microscopy and Laboratory of Microanalyses/UFMG, and Laboratório Nacional de Luz Síncrotron – Brazil

Author details

Nelcy Della Santina Mohallem[1*], Juliana Batista Silva[2], Gabriel L. Tacchi Nascimento[3] and Victor L. Guimarães[1]

*Address all correspondence to: nelcy@ufmg.br

1 ICEx – Universidade Federal de Minas Gerais, Brazil

2 Centro de Desenvolvimento de Tecnologia Nuclear - CNEN/CDTN, Brazil

3 Nanum Nanotechnology Ltda, Brazil

References

[1] Moore, J. H., & Spencer, N. D. (2001). Ed. Encyclopedia of Chemical Physics and Physical Chemistry. Taylor & Francis

[2] Brushar, B. (2004). Handbook of Nanotechnology. New York, Springer.

[3] Valenzuela, R. (1994). Magnetic Ceramics. Cambridge, Cambridge University Press, 10.1017/CBO9780511600296.

[4] Coey, J. M. (2001). Magnetism in future. *Journal of Magnetism and Magnetic Materials.*, 226-230, 2107-2112.

[5] Snelling, E. C., & Giles, A. D. (1986). Ferrites for Inductors and Transformers. 2 a. ed. NY,

[6] Kingery, W. D., Bowen, H. K., & Uhlmann, D. R. (1976). Introduction to Ceramics 2. ed. New York, John Wiley & Sons.

[7] Rajaram, R. R., & Sermon, P. A. (1985). Adsorption and catalytic properties of $Co_xFe_{3-x}O_4$ spinels. *J. Chem. Soc., Faraday Trans.*, 81, 2577-2591.

[8] Ramankutty, C. G., & Sugunan, S. (2001). Surface properties and catalytic activity of ferrospinels of nickel, cobalt and copper, prepared by soft chemical methods. *Applied Catalysis A: General.*, 212, 39-51.

[9] Khalf-Alla, M., El -Salaam, A. B. D., Said, A. A., Hassan, E. A., & El -Wahab, abd. M. M. (1994). Structure and electronic effects of cobalt ferrites, CoxFe3-xo4 on catalytic decomposition of isopropyl alcohol. *Collect. Czech. Chem. Commun.*, 59, 1939-1949.

[10] Sharifi, H., & Shokrollahi, S. A. (2012). Ferrite-based magnetic nanofluids used in hyperthermia applications Ibrahim. *J. Magnet. Magnet. Mat.*, 324, 903-915.

[11] Bocanegra-Diaz, A., Mohallem, N. D. S., & Sinisterra, R. D. (2003). Preparation of a ferrofluid using cyclodextrin and magnetite. *J. Braz. Chem. Soc.*, 14, 936-941.

[12] Silva, J. B., Brito, W. ., & Mohallem, N. D. S. (2004). Influence of the heat treatment on cobalt ferrite nanocrystal powders. *Materials Science and Engineering. B, Solid State Materials for Advanced Technology*, 112, 182-187.

[13] Hench, L. L., & West, J. K. (1990). Principles of Electronic Ceramics. Wiley.

[14] Grigorova, M., Blythe, H. J., & Blaskov, V. (1998). Magnetic properties and Mössbauer spectra of nanosized $CoFe_2O_4$ powders. *J. Magnet. Magnet. Mat*, 183, 163-172.

[15] Qu, Y., et al. (2006). The effect of reaction temperature on the particle size, structure and magnetic properties of coprecipitated $CoFe_2O_4$ nanoparticles. *Mater. Lett*, 60, 3548-3552.

[16] Song, Q., & Zhang, Z. J. (2006). Correlation between spin-orbital coupling and the superparamagnetic properties in magnetite and cobalt ferrite spinel nanocrystals. *J. Phys. Chem. B. V. 110*, 23, 11205-11209.

[17] Chinnasamy, C. N., et al. (2003). Unusually high coercivity and critical single-domain size of nearly monodispersed $CoFe_2O_4$ nanoparticles. *App. Phys. Lett.*, 83.

[18] Haneda, K., & Morrish, A. H. (1981). Magnetic structure of small NiFe2O4 particles. *J. Appl. Phys.*, 52.

[19] Haneda, K., & Morrish, A. H. (1988). Noncollinear magnetic structure of CoFe2O4 small particles. *J. Appl. Phys.*, 63.

[20] Silva, J. B. ., & Mohallem, N. D. S. (2001). Preparation of composites of ferrites dispersed in a silica matrix. *J.Magnet. Magnet. Mat*, 226, 1393-1396.

[21] Casula, M. F., Corrias, A., & Paschina, G. (2000). Nickel oxide-silica and nickel-silica aerogel and xerogel nanocomposite materials. *Journal Materials Reserch.*, 15, 2187-2194.

[22] Casula, M. F., Corrias, A., & Paschina, G. (2001). Iron oxide-silica aerogel and xerogel nanocomposite materials. *J Non- Crystall. Solids.*, 293-295, 25-31.

[23] Estounes, C., Lutz, T., & Happich, J. (1997). Nickel nanoparticles in silica gel: preparation and magnetic properties. *J. Magnet. Magnet. Mat.*, 173, 83-92.

[24] Huang, X. H., & Chen, Z. H. (2004). Sol-gel preparation and characterization of CoFe2O4-SiO_2 nanocomposites. *Solid State Commun.*, 132, 845-850.

[25] Kasemann, R., & Schmitd, H. K. (1994). A new type of a sol-gel derived inorganic-organic nanocomposite. *Materials Research Society.*, 346, 915-921.

[26] Silva, J. B. ., Diniz, C. F. ., Ardisson, J. D. ., Persiano, A., & Mohallem, N. D. S. (2004). Cobalt ferrite dispersed in a silica matrix prepared by sol-gel process. *Journal of Magnetism and Magnetic Materials*, 272-76, 1851-1853.

[27] Silva, J. B., Mohallem, N. D. S., Alburquerque, A. S., Ardisson, J. D. ., Novak, M. A., & Sinnecker, E. (2009). Magnetic studies of cofe2o4/sio2 aerogel and xerogel nanocomposites. *Journal of Nanoscience and Nanotechnology*, 9, 1-8.

[28] Souza, K. C. ., Mohallem, N. D. S. ., & Sousa, E. M. B. (2010). Mesoporous silica-magnetite nanocomposite: facile synthesis route for application in hyperthermia. *Journal of Sol-Gel Science and Technology*, 53, 418-427.

[29] Xianghui, H., & Zhenhua, C. (2006). Preparation and characterization of $CoFe_2O_4/SiO_2$ nanocomposites. *Chin. Sci. Bull.*, 51(20), 2529-2534.

[30] Xiao, S. H., Luob, K., & Zhanga, L. (2010). The structural and magnetic properties of cobalt ferrite nanoparticles formed in situ in silica matrix. *Materials Chemistry and Physics*, 123-385.

[31] Julian, C., Alcazar, G. A., & Montero, M. I. (1999). Mössbauer analysis od the phase distribution present in nanoparticulate Fe/SiO_2 samples. *Journal of Magnetism and Magnetic Materials.*, 203, 175-177.

[32] Rohilla, S., Kumar, S., Aghamkar, P., Sunder, S., & Agarwal, A. (2011). Investigations on structural and magnetic properties of cobalt ferrite/silica nanocomposites prepared by the coprecipitation method. *Journal of Magnetism and Magnetic Materials*, 323-897.

[33] Gharagozlou, M. (2010). Study on the influence of annealing temperature and ferrite content on the structural and magnetic properties of $x(NiFe_2O_4)/(100-x)SiO_2$ nanocomposites. *Journal of Alloys and Compounds*, 495-217.

[34] Gan, Y. X. (2012). Structural assessment of nanocomposites. *Micron*, 43-782.

[35] Sivakumar, N., Narayanasamy, A., Chinnasamy, C. N., & Jeyadevan, B. (2007). Electrical and magnetic properties of chemically derived nanocrystalline cobalt ferrite. *J. App. Phys.*, 102.

[36] Rajendran, M., Pullar, R. C., et al. (2001). Magnetic properties of nonocystaline $CoFe_2O_4$ powders prepared at room temperature: variation with crystallite size. *Journal of Magnetism and Magnetic Materials.*, 232, 71-83.

[37] George, M., et al. (2007). Finite size effects on the electrical properties of sol-gel synthesized $CoFe_2O_4$ powders: deviation from Maxwell-Wagner theory and evidence of surface polarization effects. *J. Phys. D: App. Phys.*, 40, 1593-1602.

[38] Meyer, K., & Lorenz, P. (1994). Porous solids and their characterization. *Cryst. Res. Technol.*, 29, 903-930.

[39] Lowell, S., & Shields, J. E. (1984). Power surface area and porosity. 2 a ed.,John Wiley N. Y.

[40] Gelb, L., & Gubbins, K. E. (1998). Characterization of porous glasses. *Langmuir*, 14, 2097-2111.

[41] Gregg, S. J., & Sing, K. S. W. (1982). Adsorption, surface area and porosity. 2 a ed. NY,

[42] Aegerter, M. A. (1989). Sol-Gel Science and Technology. .London World Scientific Publishing

[43] Brinker, C. J., & Scherer, G. W. (1990). Sol-gel science: The physics and chemistry of sol-gel processing. Academic Press Inc.

[44] Livage, J., Henry, M., & Sanchez, C. (1988). Sol-gel chemistry of transition metal oxides. *Prog. Solid State Chem.*, 18.

[45] Klein, E. . (1994). Sol-Gel Optics: processing and applications. Kluwer Academic Publishers.

[46] Mac, Kenzie. J. D. (2003). Sol-gel research- achievements since 1981 and prospects for the future. *J. Sol-Gel Sci. Techn.*, 23-27.

[47] Panjonk, G. M., Venkateswara, A., & Sawant, B. M. (1997). Dependence of monolithicity and physical properties of TMOS silica aerogels on gel aging and drying conditions. *Journal Non-Crystalline Solids.*, 209.

[48] Mohallem, N. D. S., Santos, D. I., & Aegerter, M. A. (1986). Fabricação de sílica vítrea pelo método sol-gel. Cerâmica, , 32(197), 109-116.

[49] Burneau, A., Gallas, J. P., & Lavalley, J. C. (1990). Comparative study of surface hidroxyl groups of fumed and precipitated silicas. *Langmuir*, 6, 1364-1372.

[50] Casu, M., & Casula, M. F. (2003). Textural characterization of high temperature silica xerogels. *Journal of Non-Crystalline Solids.*, 315, 97-106.

[51] Rao, A. V., Haranath, P. B. W., & Risbud, P. P. (1999). Influence of temperature on the physical properties of TEOS silica xerogels. *Ceramics International.*, 25, 505-509.

[52] Stolarski, M., & Pniak, M. S. B. (1999). Synthesis and characteristic of silica aerogels. *Applied Catalysis A: General.*, 177, 139-148.

[53] Schneider, M., & Baiker, A. (1995). Aerogels in catalysis. *Catal. Ver. Sci. Eng.*, 37, 515-556.

[54] Waldron, R. D. (1955). Infrared Spectra of Ferrites. *Physical Review.*, 99(6), 1725-1732.

Chapter 8

Synthesis and Characterization of Ti-Si-C-N Nanocomposite Coatings Prepared by a Filtered Vacuum Arc with Organosilane Precursors

Seunghun Lee, P. Vijai Bharathy, T. Elangovan, Do-Geun Kim and Jong-Kuk Kim

Additional information is available at the end of the chapter

1. Introduction

Many deposition tools such as magnetron sputtering, plasma enhanced chemical vapor deposition (PECVD), arc ion plating (AIP), and filtered vacuum arc (FVA) have been introduced for synthesizing nanocomposite films. Table 1 summarizes previous works of nanocomposite coatings. Nanocomposites based on TiN have been investigated dominantly with the incorporation of silicon or carbon contents. The incorporation methods such as an alloy arc cathode, addition of reactive gas, and additional magnetron sputtering have been used to deposit ternary or quaternary composition nanocomposite films. The magnetron sputtering and PECVD have been firstly used to grow nanocomposite films due to the simplicity of controlling a composition ratio becuase precise control of additional components is important to make a nanocomposite structure. For example, Ti-Si-N nanocomposite films have showed the maximum hardness at Si content of 9±1 at.% [1]. After that, a vacuum arc discharge has been applied to the nanocomposite coatings becuase the vacuum arc process has many advantages against other CVD or PVD processes. A vacuum arc plasma exhibits high ionization ratio more than 90%. Also the ion energy in a vacuum arc is in the range of 10-100 eV. Hence, the effect of ion energy on the film structure appears significantly [2]. Nevertheless, the vacuum arc method cannot avoid the problem of macro particles. Macro particles generated from arc spots are the main drawback of the vacuum arc process. The macro particles form micro cracks or pin holes, resulting in a bad corrosion resistance when coatings are exposed in some corrosive environments.

In FVA, magnetic filters have been introduced to transport plasma except the macro particles. The filters transport charged particles selectively using electromagnetic fields. However, neutral macro particles collide with a wall by an inertia drift. Main issues are the efficient removal of the macro particles and the minimization of ion loss through a magnetic filter wall. The effective way to reduce the macro particles is based on the spatial separation of the trajectories of macro particles and ions [3]. If the magnetic field is curved such as the field inside a curved solenoid, electrons follow the curvature. The electrons are said to be magnetized. In contrast, ions are usually not magnetized because the gyration radius of ion is much larger than that of electron. Nevertheless the ions are forced to follow the magnetic field lines due to the electric fields between electrons and ions. Therefore ions and electrons are transported along magnetic field lines [4]. Various FVA methods have been widely applied to the nanocomposite deposition without any macro particle problems. Magnetic filtering technology removes efficiently the macro particles and result in a smooth film surface [5,6].

Method	Material	Hardness	Ref.
Arc	Ti-Si-N	45 GPa	[1]
Arc	Ti-Al-N/Cr-N	37 GPa	[7]
Arc	Ti-Al-Si-N	34 GPa, 42.4 GPa	[8,9]
Arc	Ti-Al-N	35.5 GPa	[10]
Arc, magnetron sputter	TiN-Cu, CrN-Cu, MoN-Cu	27-42 GPa	[11]
Arc, magnetron sputter	Ti-Si-N	45-55 GPa	[12,13]
FVA, magnetron sputter	Ti-Si-N	45 GPa	[14]
FVA, magnetron sputter, E-beam evaporation	Ti-Cr-N,Ti-B-C	43.2 GPa	[15,16]
FVA	Ti-Si-N	40.1 GPa,	[17]
PECVD	Ti-Si-N	3500 HK(kg/mm²), 40 GPa	[18,19]
PECVD	Ti-Si-C-N	48 GPa	[20]
PECVD	Ti-Si-C-N	52 GPa	[21]
Magnetron sputter	Ti-Si-N	38GPa, 45 GPa	[22,23]

Table 1. Nanocomposite coatings by various methods.

Several studies found the maximum efficiency and the optimum condition for the curved magnetic filters. An analysis of plasma motion along the toroidal magnetic field has been shown that plasma that is transported by the magnetic field in a guiding duct should satisfy the following relation [24],

$$B > M_i V_3 / Zea \tag{1}$$

where M_i is the ion mass, V_o is the translational velocity, Z is the charge multiplicity of the ion, e is the electron charge, and a is the minor radius of the plasma guiding duct. The transport of heavy metal ions having an energy of even a few tens of eV requires strong magnetic field above 1 Tesla to fulfill the inequality in Eq. (1). However, it is practically impossible to provide a stable burning of the direct current arc discharge in the strong magnetic field. Therefore, it is reasonable to consider heavy-element plasma flow transport in a curvilinear system with crossed electric and magnetic fields using the principles of plasma optics as a guide [25,26]. In this case, the required magnetic field is determined by the following condition, $\rho_e < a < \rho_i$, where ρ_e and ρ_i are the electron and ion Larmor radius, respectively. The required field is significantly lower than the fields defined by the expression in Eq. (1). Electron Larmor radius is

$$r_e = \left(m_e k T_e\right)^{1/2} / \ eB \tag{2}$$

where m_e is the electron mass, k is Boltzmann constant, and T_e is the electron temperature.

Note that electrons are only magnetized, while the ions are not. The electrons move along the magnetic field lines. Due to the highly conductive plasma, the magnetic field lines are equi-potentials. Considering a plasma diffusion in vacuum, electrons have higher mobility than ions due to smaller mass except at a sheath boundary. However, electrons expand with the same velocity as ions because electrostatic forces keep the electrons and ions together. And a cross field diffusion is given by the Bohm formula, $D_B = kT_e/16B$, though the cross field diffusion coefficient, D, is proportional to B^{-2} in the classical theory[27].

Anders mentioned about the criterion of system efficient, K_s, which is generally considered as the ratio of the total ion flow at the exit of the system, I_i, to the arc discharge current, I_a, as follows,

$$K_s = I_i / I_a \tag{3}$$

The system coefficient is typically 1% [6]. There is a general agreement that the transport efficiency is maximized by focusing the plasma into the duct and biasing the duct to a positive potential of 20 V. Predictions of the available maximum transmission vary between 11 and 25% depending on the ion energy [28]. In practice, the transport of plasma produced by pulsed high current arcs (HCA) was showed that the system coefficient was 7% [6]. For linear FVA, the maximum value of the system efficiency reached 8% when arc current Ia was adjusted in the range of 100–110 A and the magnetic filter field was ~ 20 mT [29].

To supplement the accuracy of system coefficient, a particle system coefficient, K_p, is proposed to eliminate the influence of the various ion charge states by considering the mean ion charge state, Z_{av}, of the used metal as follows.

$$K_p = I_i / Z_{av} I_a \tag{4}$$

Because the average ion charge state is taken into account, the particle system coefficient is more closely related to the deposition rate [30]. However this can be particularly insufficient for filter optimization when the system used a graphite cathode that generates solid rebounding macro particles.

The problem has been solved by the numerical calculation of the particle trajectories using a two-dimensional approximation [31]. In the calculation, it assumes that the macro particles are solid spheres, the inner surfaces of the plasma guide, and the intercepting fins are smooth, the repulsion of particles from a guiding duct wall is partially elastic, the particles are emitted froma cathode spot with equi-probability in any direction. The computing results make it possible to estimate the ratio of the pass of macro particle flow,N_{ex} to the flow, N_{ent} generated by the cathode spot. The ratio N_{ex}/N_{ent} characterizes the likelihood of an macro particle passing through the system. The results of simulations indicate that the absence of a direct line-of-sight between the cathode and the substrate is not always sufficient to provide the required degree of macro particle removal from the plasma. The results of computations performed for various magnetic filters are presented in Table 2 [27].

Filter type	Knee (45°)	Torus (45°)	Rectang.	Dome	Torus (90°)	Retil.	Radial	Wide apert.
N_{ex}/N_{ent} [%] (predicted)	1.7	25.0	17.0	1.7	0	4.4	0	0
Transport [%] (measured)	3.0	2.5	2.5	2.5	1.5	1.8	8.4	~6.0

Table 2. Fitlering (N_{ex}/N_{ent}) and transporting properties of magnetic plasma filters [27].

There are various types of filters as shown in Fig. 1[32]. Most types are used magnetic fields to transport plasma without macro particles. Several types only use the collisional reduction of macro particles. In most cases a plasma is transported from the cathode to the substrate, and the droplets are eliminated by a plasma transportation wall, guiding duct. Many review papers of the filtered arc system and technology have been reported [6,33-38]. A typical filtered arc system with its different electromagnetic plasma transportation duct or droplet filter configurations is shown in Fig. 1(a)–(h). Electromagnetic coils transporting plasma in the out of line of sight direction can be positioned in the chamber, instead of placing them outside of the filter duct. The off-plane double bend filter is nicknamed FCVA and is now commercially available [39,40]. Most FAD units have electromagnetic coils outside of the plasma duct and have baffles inside the duct wall. However, some types have freestanding coils inside the plasma duct or the chamber. Other interesting filters have been developed. Examples are shown in Fig. 1(i)–(l). In the Venetian-blind filter, the plasma passes between the vane lamellae, and the droplets are caught or reflected by the lamellae [41,42]. A coaxial filter is operated with a large current pulse, and the plasma is driven by a self magnetic field [43]. An electrostatic filter can be used with a pulsed arc having a laser trigger [44]. However, recently only the laser triggered arc is used without the electrostatic filter. Mechanical filters can be used in a pulse arc [45], which may also be used in pulsed laser deposition [46].

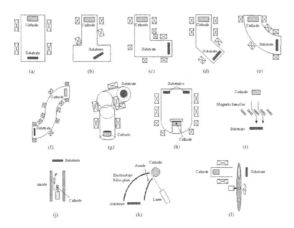

Figure 1. Various types of filter systems [32]. (a) Rectilinear. (b) Bent. (c) Rectangular. (d) Knee. (e) Torus. (f) S-shape. (g) Off-plane double bend. (h) Dome. (i) Venetian blind. (j) Co-axial (pulse). (k) Electrostatic filter with laser trigger (pulse). (l) Mechanical pulse.

2. Nanocomposite Films Prepared by a FVA With an Organosilane Precursor

2.1. Basic Configuration of Deposition System for an Organosilane Incorporated FVA

Ti-Si-C-N quaternary nanocomposite coatings were prepared by using a filtered vacuum arc deposition system. Figure 2 represents the schematic diagram of the FVA coating system. The deposition system consists of a water-cooled cathode and anode, plasma guiding duct, and magnet coils. The diameter and height of the chamber is 800 and 650 mm, respectively. The vacuum arc discharge from cathode emits high energy (~ 60 eV) ions and generates dense plasma above 10^{13} cm^{-3}. An arc spot also makes neutral macro particles, which cause several problems such as rough surface, pinhole, and micro cracks in coatings. To remove the macro particles the plasma guiding duct and magnet coil were used. Coaxial magnetic fields about 15 mT were generated by six magnetic coils. The pumping system consists of rotary pump (900 l/min) and oil diffusion pump (1500 l/s). The ultimate pressure of deposition chamber was 2×10^{-6} mTorr by using rotary and diffusion pumps. MKS mass flow controllers were used to regulate the flow rate of tetramethylsilane (TMS) (99.99%), argon (99.999%) and nitrogen (99.99%) gases.

The samples are mounted on the rotational substrate holder with a rotational speed of 3 rpm for the deposition of the coatings. The deposition process consisted of Ar ion bombardment cleaning, Ti(100 nm)/TiN(200 nm) interlayer deposition to improve an adhesion strength. The thickness of the Ti-Si-C-N nanocomposite coating was about 0.4 to 0.6 μm. The experimetal details are shown in Table 3.

Structural characterization of as-obtained samples was done by X-ray diffraction (Shimadzu XRD-6000, Cu Kα radiation λ = 1.5406 Å, scanning rate 1° min−1) and transmission electron microscopy (JEOL JEM-3100 FEF-UHR, 300 kV). The crystallite size was determined by Scherrer formula and lattice parameter of coating was calculated by Bragg law for the cubic system.

Elemental analysis and chemical nature of coatings were performed with X-ray photoelectron spectroscopy (XPS), using a VG Scientific ESCALAB 250 spectrometer with a Mg Kα X-ray source. Hardness are assessed by means of a nanoindentation system (MTS, nano indenter XP) using a Berkovich diamond indenter. Coating hardness was determined from the loading and unloading curves employing depth-sensing hardness testers. The applied load was gradually increased to 3000 μN at a loading rate of 150 μN/s, and was held at this maximum value for 10 s. Testing was done using the constant-displacement-rate mode until a depth of 200 nm was reached and the values from fifteen indents were averaged for each condition. Adhesion strength was measured by a scratch tester (J&L, Scratch Tester). The applied load on the diamond tip (tip radius 200 μm, conical angle 120°) was continuously increased at a rate of 0.25 N/s, while the tip advanced at a constant speed of 0.05 mm/s.

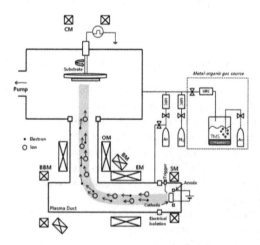

Figure 2. Experimental schematics of filtered vacuum arc with organosilane vaporizations.

Arc voltage	24 V
Arc current	60 A
Duct voltage	25 V
Duct current	43A
Substrate voltage	0 ~ -400 V
Base pressure	2.3×10^{-6} Torr
Total pressure	8.2×10^{-4} Torr
Deposition rate	9-38 nm/min
N2 flow rate	50 sccm
TMS flow rate	10 sccm
Substrate Temperature	< 150°C

Table 3. Experiment details.

2.2. Ti-Si-C-N Nanocomposite Film Prepared by a FVA With TMS Gas

2.2.1. Composition and Chemical Analysis

XPS analysis is used to identify a chemical composition on the surface of the nanocomposite coating. Surface pre-sputtering eliminated the surface oxide at air-exposed samples. The elemental compositions for Ti, Si, C and N as a function of TMS flow rate from 5 to 20 sccm are given in Fig. 3. A strong increase of the Si (2.1±1 to 12.2±1 at. %) content is observed up to the TMS flow rate of 20 sccm. Ti and N content showed decreasing trend from 18.8+1 to 9.83±1 at.% and 24.8±2 to 19.14±2 at. %, respectively. Whereas the carbon content remained constant in the range of 31± 2 at. %.

The XPS spectra of of Ti 2p, Si 2p, and N 1s are shown in Fig. 4. Fig. 4 (a) depicts the XPS spectrum of Ti 2p. The characteristic doublet of Ti $2p_{3/2}$ and $2p_{1/2}$ are clearly observed. The Ti $2p_{3/2}$ peak position shows positive shift from 460.85 eV, which reveals a good agreement value of 460.5 eV for TiN. This peak position shifts to higher value at silicon content of 3 at. % and then it moves to lower binding energy (BE) side at silicon content at 8 at. %. The second peak position Ti $2p_{1/2}$ is showing at 455.15 eV matching with Ti- nature bonding position of TiN. It is well known that TiN_x has flexible chemical states governed by the compactness of other small non-metal atoms filled in the octahedral voids of titanium [47]. Fig 4 (b) shows the high resolution spectrum of N 1s region. The three peaks correspond to 398.5, 396.8 and 400.6 eV, are in agreeing with the binding energies of Si_3N_4, TiN and TiC, respectively. With increasing silicon content the N1s peak position showeda negative shift from 396.8 to 396.6 eV. Fig 4 (c) shows Si 2p spectra peaks, which are observed at 101.8, 101.1 and 102.8 eV, are attributed to the Si_3N_4, the Si (2p)–C, and Si (2p)–O bonds, respectively. The incorporation of TMS at low Si content (2 at.%) results in Si_3N_4 formation, whereas SiC formation is dominant at the high Si content (12 at.%). The reason is that CH_3 in TMS is not dissociated perfectly in arc plasma so the carbon content are incorporated and react with Si contents. From the XPS results, it was concluded that Si in Ti–Si–C–N coatings existed mainly as amorphous silicon nitride with some silicon carbide.

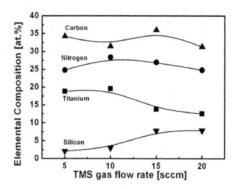

Figure 3. Chemical composition of Ti-Si-C-N nanocomposite films as a function of TMS flow rate.

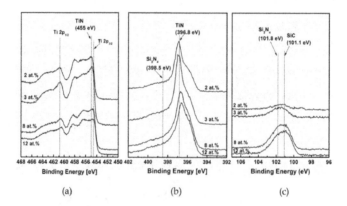

(a)   (b)   (c)

Figure 4. XPS spectra of Ti-Si-C-N nanocomposite films. (a) Ti (2p), (b) Si (2p),(c) N (1s).

2.2.2. Microstructure Analysis

X-ray diffraction (XRD) is used to investigate the crystalline phases of the Ti-Si-C-N coating.
Figure 5 shows the XRD pattern of Ti-Si-C-N coating with different silicon contents. The Ti-
Si-C-N nanocomposite coatings with silicon content range of 2.1 to 16.2 at. %, exhibit the diffraction peaks at the angles of $2\theta = 36.01°$, $43.63°$ and $50.79°$ which are corresponding to
(111), (200) and 220) TiCN reflections respectively [14,48-52]. These positions of the three
peaks are coinciding with the values obtained in the JCPDS card [53]. Also, peak is observed
at $38°$ that are attributed to the diffraction of stainless steel, which is the substrate. No signals from the phase formation of Si_3N_4 or from titanium silicide can be observed [54,55].
Note that the amorphous phase Si_3N_4, deduced from the XPS results analysis, has been confirmed with the XRD measurement.

The crystalline size of the Ti-Si-C-N coating is calculated from the TiN (111) diffraction peak. The TiN (111) peak is fitted by using a Gaussian function to calculate the crystallite size from FWHM by using Scherrer equation. Adding Si, the crystallite sizes decreases from 3 to 2 nm, which shows that nanocrystalline phases are formed. It representsthat the addition of silicon could reduce the grain coarsening in the Ti-S-C-N nanocomposite coatings.

Figure 6 shows HRTEM picture acquired on the Ti-Si-C-N coating contain 3 at. % of Si. The lattice spacing of these crystallites is 0.212 nm. The amorphous phase has an irregular shape and a boundary surrounding the TiN nanocrystallites. The growth direction of (111) plane is clearly identified in this image. The image shows the fine grain structure and reveals that these fine grains are largely oriented in the direction of growth. An arrow indicated the growth direction. TiN nanocrystals with an average grain size of about 10 nm were separated by less than 1 nm thick brighter Si_3N_4 tissues. Fig. 6 (b) shows a STEM image of Ti-Si-C-N coating and the intensities from Ti, Si, C, and N obtained from an EDX line scan acquired from the STEM image. In the matrix, the Ti and the Ti signals are high and small for silicon small signals are observed from the XRD, XPS and TEM analysis results, it could be confirmed that the Ti–Si–C–N coating obtained in this experiment consisted of nanosized TiN crystallites surrounded by thin amorphous phase of Si_3N_4 [56].

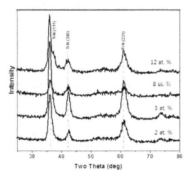

Figure 5. X-ray diffraction patterns at various Si contents from 2 to 12 at. %.

Figure 6. HRTEM image of Ti-Si-C-N at 3 at.% of Si content.

2.2.3. Mechanical Properties

The nano-indentation technique permits to extract the surface mechanical properties from depths of nanometers. Hardness can be calculated using Oliver and Pharr method [57]. The hardness decreases with increasing the depth of indentation at extremely small depths. This indentation sizes effect is expected for soft metal films and has been related to strain gradient plasticity. In the present study, nano-hardness of the Ti-Si-C-N coatings was obtained as a function of depth up to a maximum depth of 180 nm. The steel substrate hardness is around 2 GPa and Young's modulus is 200 GPa as obtained by nano-indentation experiments. Load versus indentation depth curves from multiple experiments by using the same maximum load and from 25 different sample locations were averaged and standard deviations were calculated and reported. Figure 7 represents the hardness of Ti-Si-C-N coatings as a function of displacement into surface. The hardness values of Ti-Si-C-N coatings are 18, 35, 32 and 27 GPa at a contact depth of 40 nm at 2.1, 3, 7 and 8 at. % of Si, respectively. The maximum hardness of ~35 GPa is obtained for the silicon content of 3 at. %. The hardness value of 48 GPa was achieved adding 10 at.% Si and 30 at.% C by Dayan Ma et.al [58]. Also he was explained it with carbon contents. Suddeep mabiraham et. al [49] also achieved the hardness of the Ti-Si-C-N films at the Si content of 9.2. at % with maxiumum hardness around 55 GPa achieved with Si content of 8.9 at.%. by using PVD method in [59]. So far many people have been reported the hardness of the Ti-Si-C-N coating at above 5 at. % of the silicon content. Present results show that the maximum increases in hardness for Ti-Si-C-N coating within 3 at. % of silicon, which proves such hardening effects, is known to occur in transition metal nitride systems with a few at. % of silicon content and this increased hardness effect due to hindrance effect of the segregated Si_3N_4 on TiN grain boundary sliding which is the predominant deformation mechanism in nanocrystalline materials [60]. On the other hand, the hardness reduction with a further increase in Si content is observed [58]. The decrease in hardness is due to increasing contributions from the soft substrate.

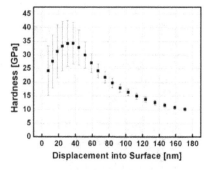

Figure 7. Nano indentation of Ti-Si-C-N films with 3 at. % of Si.

2.2.4. Adhesion Properties

Scratch tests with ramping loads up to 100 N were conducted on the films deposited on stainless steel substrates. The critical load required to cause the first delamination at the edge of the scratch track (adhesive failure) characterizes the adhesion properties of coatings. Optical microscopy of the films subjected to the scratch tests (Not shown here). The critical loads obtained for the above four coatings were 37, 44, 33 and 30 N, respectively. Strong acoustic signals were observed for the critical load exceeding 30 N. The lowest value of COF of 0.16 is obtained for the hardest film with Si content of 2.1 at.% and it is 0.17- 0.22 for the rest of nanocomposite films and agrees with the literature [59, 61]. For FVA deposition, the intrinsic energy of ions emitted from a cathode source is considerably higher as compared to the evaporated or sputtered atoms. This higher ion energy up to 60 eV condenses deposited film and enhances adhesion. Beside that the friction coefficient and penetration depth versus force show several inflections for the Ti-Si-C-N with 8 at. % Si.

2.3. Substrate Bias Effects on Ti-Si-C-N Nanocomposite Films

2.3.1. Structural Properties

The deposition rate of Ti-Si-C-N nanocomposite coating was found to decrease gradually from 18 to 8 nm/min, when the bias voltage increased from 0 to −400 V. The decreasing trend of deposition rate has been explained on the basis of removal of impurities and densification of films due to the energetic ion bombardment when increased substrate bias voltage [62]. The XRD diffraction of Ti-Si-C-N nanocomposite coating for different substrate bias voltage is as shown in Fig. 8. It matches well with the ICDD PDF 42-1489 revealing the NaCl type crystal structure. However, the diffraction peaks are shifted towards the higher angle side indicating the presence of compressive stress in the coatings [63]. At zero bias voltage, coating shows predominant presence of (111) and (200) orientations. With increase in substrate bias, (220) was found to dominate the spectra especially at substrate biases of − 300 V and − 400 V. This behavior has been observed in binary, ternary and quaternary coatings and is attributed to the decrease in sputtering rate at higher substrate bias. There is no further detail of the Ti-Si-C-N coating with effect of bias voltage by using FVA system. These result show that substrate bias voltage has a strong influence on the structural properties of the deposited films and will be correlated with changes in hardness of the coatings. Fig. 9 shows the calculated lattice parameters from the XRD pattern for TiN peak position using equation (1). The lattice constant of TiN changes from 4.31599 to 4.24762 Å with the different substrate bias voltages. For −100 V bias voltage, TiN films are over stoichiometric corresponding to the highest lattice parameter (ICDD PDF 42-1489) a =4.29 Å). The shift could be attributed to higher residual stress in the coatings and changes in the composition of the coating. The calculated lattice parameter was shown 4.24762 Å at −400 V, it is slightly lower than that of the value reported standard lattice parameter. Usually these kinds of stress with increasing bias voltage are commonly observed in thin films grown by physical vapour deposition methods. With increasing bias voltage, the ion bombardment encouraged mobility of atoms was major effect which implies that the decreasing trend of the stress because of enhanced annihilation of defects [64].

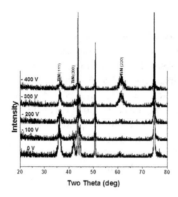

Figure 8. X-ray diffraction patterns of Ti-Si-C-N nanocomposite coatings various substrate bias voltage.

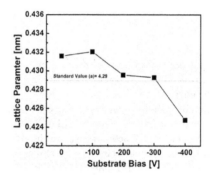

Figure 9. Lattice parameter of the Ti-Si-C-N nanocomposite coatings various substrate bias voltage.

2.3.2. Texture Orientation

The changes in the preferred orientation of Ti-Si-C-N nanocomposite coating as a function of the substrate bias are qualitatively estimated in terms of texture coefficients (TC). The TCs, determined by TC = $I_m(hkl)/I_0(hkl)/(1/n)\{I_m(hkl)/I_0(hkl)\}$ for (111) and (220) reflection [65], as a function of substrate bias are shown in Fig 10. Where $I_m(hkl)$ is the observed intensity of the (hkl) plane, $I_0(hkl)$ is the standard data (JCPDS) of the (hkl) plane, and N is the total number of diffraction peaks. When the TC value is larger than 1, a preferred orientation exists in the sample. The texture coefficient of the (111) orientation is significantly higher than that of the (220) orientation in the substrate bias range 0 to −100 V. When deposited lower film thickness it has been shown that the influence of surface prevails over strain energy and (100) orientation expected, but it is higher film thickness have (111) orientation expected with vice versa [66]. However we observed the contrary result compared to other researchers for nitride coatings with substrate bias. This may be due to the ion input energy or channeling effect is not the major role for the change of orientation between (111) and (220) plane. At −300 V, the trend of texture has completely changed from (111) to (220) and texture coefficient value shows extensively higher value than that of the (111) and it was maintained at the higher bias voltage of -400 V. It attributed to the mutual effects of both Ti and Si elements for preferred growth orientation of the Ti-Si-C-N nanocomposite coating at higher substrate bias voltage.

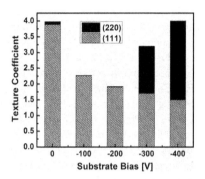

Figure 10. Texture Coefficient of the Ti-Si-C-N nanocomposite coatings various substrate bias voltage.

2.3.3. Chemical Composition and Analysis

To determine the composition of the Ti-Si-C-N nanocomposite coating, XPS analysis has been performed. The samples are sputtered with argon ions to remove the oxide top layer. Even though we found uniform oxygen concentration with respect to substrate bias voltage, could be due to the TMS gas which is used in the present studies. With increasing the negative bias voltage from 0 to −300 V, the silicon concentration increased from 5.17 to 8.14 at. %. Silicon has lower electro negativity than carbon; amorphous carbon atom bonded to a silicon atom attracts electrons from the silicon atom, which condense the Si-H bonds. The number of silicon atoms attached to Si-Hx is predicted to increase as silicon content increases in the Ti-Si-C-N nanocomposite coating [67]. In contrast, decreasing trend was observed for carbon content in the range from 40 to 24 at. %. It may be due to carbon atom is lighter than the Si atom and also the energy of the Ar ion increases with cause to build strong bombarding to the substrate with increasing bias voltage. Further with increasing substrate bias voltage about −400V, the silicon concentration shows decreasing trend in the range of 4 at.%, it is also lower value than substrate bias voltage range of silicon content from 0 to -300 V. The N and Ti content are seen to be almost independent respect to the substrate bias.

The chemical state of the Ti-Si-C-N nanocomposite coating was analyzed by the XPS measurements and is as depicted in Fig. 11. Fig 11 (a) depicts the C 1s peak spectra shows five component's namely, C1, C2, C3, C4 and C5 were observed in the spectrum zone. The two peaks of C1 (at 285.7 eV) and C2 (at 284.5 eV) corresponded with the position of C-N and C=C phase formation. Another three peak were peaks observed C3 (at 282.5 eV), C4 (at 281.2 eV) C5 (at

288.4 eV) corresponded with the position of Ti-C phase formation. As compared Ti-C peak position, the C=C peak position accounts for only a small fraction of the total C1 spectra, indicating that a small fraction of C atoms are bonded to Ti atoms, and most of the C atoms exist as amorphous carbon [68,69]. Fig 11(b) depicts the Si 2p peak spectra shows three components namely S1, S2 and S3 were observed in the spectrum zone. These two peaks of S1 (at 100.8 eV) and S2 (at 102.3 eV) corresponded with the position of SiC and Si_3N_4 phase formation. Another one of weak component was observed at about S3 (at 102.8 eV) corresponds to Si-O bonds. The peak intensity gradually increased with an increase of bias voltage, which implies that increasing trend of silicon concentration. Fig. 11(c) depicts the N1s 2p peak spectra shows three components namely N1, N2 and N3 were observed in the spectrum zone. These two peaks of N1 (at 398.28 eV) and N2 (at 396.64 eV) corresponded with the position of Si_3N_4 and TiN phase formation. Another one of the weak component at about N3 (at 400.8 eV) corresponds to C-N bonds. The formation of Si_3N_4 phases was confirmed by XPS analyses for nanocomposite coatings. Figure 12 shows the HRTEM image of Ti-Si-C-N nanocomposite coatings deposited at a bias voltage of −100 V. These coatings are nanocomposite coatings of TiN nano-crystalline (black area) embedded in an amorphous matrix, which are clearly distinguished from the particles by high-resolution TEM image. The SAED patterns in the TEM analysis did not reveal any crystalline silicon nitride. Finally from the XRD, TEM and XPS result together confirmed that the Ti-Si-C-N coatings had nanocomposites structure of nanosized TiN crystallites embedded in an amorphous Si_3N_4 matrix.

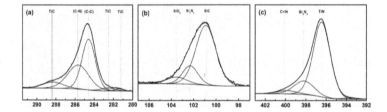

Figure 11. XPS spectra of Ti-Si-C-N nanocomposite coatings with various substrate bias voltage (a) C (b) Si and (c) N.

Figure 12. HRTEM image of Ti-Si-C-N coating at -100 V.

2.3.4. Nano Mechanical Properties

Figure 13 shows the effect of substrate biasing on the hardness of Ti-Si-C-N nanocomposite coating. It was clearly seen that the hardness increases initially with the increase in the indentation depth upto 30 nm. After that the hardness starts to decrease with increase in the penetration depth and finally attains saturation. This saturation in hardness value is observed in all the coatings. Hence this clearly indicates that at higher penetration depths, the obtained hardness may be due to the substrate effect. Hence, as per the Oliver-Pharr [70], the extracted hardness values are only from the 10 % of the film thickness. Hence, for the un-biased Ti-Si-C-N coating,at the depth of about 30 nm, the peak hardness was found to be 49 GPa. A maximum value of 49 GPa has been obtained for the coatings deposited at a substrate bias of − 100 V. This peak in hardness value is well corroborated with XRD, XPS and TEM studies. Furthermore increasing trend of hardness with bias voltage have been also studied many researchers by using different PVD synthesis methods. Usually enhancement of the packing density in plane of (111) were improved with increasing the substrate bias voltage and also from in our XRD pattern already confirms about orientation (111) has been formed at this bias range of coating. It is well known resultant microstructure of the film depends on the ion bombardment [71], and this in turn affects the hardness of the film. Also from the TEM analysis we have confirmed nanocrystalline phase formation with amorphous phase formation in this range of bias (–100 V) coating, it may be another reason for peaking hardness at this range of nanocomposite coating. Finally with increase the substrate bias the energy of the bombarding ions increases, causing structural modification, which is responsible for changes in the level of the hardness. However, a drastic drop of hardness observed from 49 to 20 GPa over the range −100 and −200 V was observed. The film hardness between −200 to −300 V substrate biases remained constant at a value of about 18.5 GPa. We already pointed out in our XRD result shows the preferred orientation of (220) and stress formation with more defects this causes decreasing hardness. Therefore we can conclude that the hardness of nanocomposite coating depends directly on the substrate bias of nanocomposite coating in this present studies.

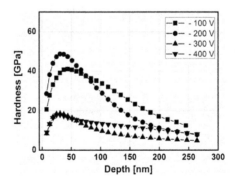

Figure 13. Nanohardness as a function of displacement into surface at various substrate voltages.

3. Summary

Ti-Si-C-N nanocomposite thin films on stainless steel were prepared by using FVA technique at constant gas mixture of argon and nitrogen flow rate with room temperature. The nanocomposite films with silicon content in the range of 2.1 to 16.2 at. % was prepared on stainless steel substrate with different TMS gas using FVA technique. From the XRD pattern, we have confirmed nancomposite structure like nc-TiCN/a-Si$_3$N$_4$ formation. The nanocrystallite size of the samples decreases with the silicon content. Nanohardness measurement indicated a peak hardness of ~49 GPa and Young's modulus of ~245 GPa for the films with Si content of 3 at.%. All these results show that Ti-Si-C-N nanocomposite coatings are suitable for surface coatings applications requiring low roughness, moderate hardness and low friction coefficient.

In the variation of substrate bias, we found (111) orientation at lower bias voltage range from 0 to – 100 V, whereas the (220) orientation has confirmed at higher bias voltage range from –200 to –400V. XPS result of Ti-Si-C-N coatings has confirmed the formation of nc-Ti(C)N/a-Si$_3$N$_4$ phase with respect bias voltage. The highest hardness around at 49 GPa has achieved phase at the bias voltage of -100 V. Further increasing voltage the hardness was decreased due to stress and orientation behavior on this coating. By changing the bias voltage to change microstructure and chemical natural of the films and to improve tribological applications such as hardness and adhesion properties, it shows promising future in industrial fields.

Author details

Seunghun Lee, P. Vijai Bharathy, T. Elangovan, Do-Geun Kim* and Jong-Kuk Kim

*Address all correspondence to: dogeunkim@kims.re.kr

Korea Institute of Materials Science, Changwon, Republic of Korea

References

[1] Yang, Sheng-Min, Chang, Yin-Yu, Wang, Da-Yung, Lin, Dong-Yih, & Wu, WeiTe. (2007). *Journal of Alloys and Compounds*, 440, 375-379.

[2] Miller, H. C. (1981). *J. Appl. Phys.*, 52, 4523.

[3] Aksenov, I. I., Strel'nitskij, V. E., & Vasilyev, V. V. (2003). *D.Yu. Zaleskij. Surf. and Coat. Technol.*, 163-164, 118-127.

[4] Sanders, D. M., & Anders, A. (2000). *Surf. Coat. Technol.*, 133/134, 78-90.

[5]   Aksenov, I. I., Vakula, S. I., Padalka, V. G., & Khoroshikh, V. M. (1980). Sov. 350 mA
      when the arc current was 60 A and the extrac-. *Phys. Tech. Phys.*, 25, 1164.

[6]   Kim, Jong-Kuk, Lee, Kwang-Ryeol, Eun, Kwang Yong, & Chung, Kie-Hyung. (2000).
      Surf. Coat. Technol., 124, 135-141.

[7]   Chang, Yin-Yu, Wang, Da-Yung, & Hung, Chi-Yung. (2005). *Surf. Coat. Technol.*, 200,
      1702-1708.

[8]   Tanaka, Y., Ichimiya, N., Onishi, Y., & Yamada, Y. (2001). *Surf. Coat. Technol.*, 146-147,
      215-221.

[9]   Li, Chen., Yong, Du., Wang, Ai J., Wang, She. Q., & Zhou, Shu. Z. (2009). *Int. Journal
      of Refractory Metals & Hard Materials*, 27, 718-721.

[10] Bujak, J., Walkowicz, J., & Kusinski, J. (2004). *Surf. Coat. Technol.*, 180-181, 150-157.

[11] Ozturk, A., Ezirmik, K. V., Kazmanl, K., Urgen, M., Eryılmaz, O. L., & Erdemir, A.
      (2008). *Tribology International*, 41, 49-59.

[12] Choi, Sung Ryong, Park, In-Wook, Kim, Sang Ho, & Kim, KwangHo. (2004). *Thin Sol-
      id Films*, 447-448, 371-376.

[13] Kim, KwangHo, Choi, Sung-ryong, & Yoon, Soon-young. (2002). *Surf. Coat. Technol*,
      298, 243-248.

[14] Kim, Do-Geun, Svadkovski, Igor, Lee, Seunghun, Choi, Jong-Won, & Kim, Jong-Kuk.
      (2009). *Current Applied Physics*, 9, S179-S181.

[15] Gorokhovsky, V. I., Bowman, C., Gannon, P. E., Van Vorous, D., Voevodin, A. A.,
      Muratore, C., Kang, Y. S., & Hu, J. J. (2008). *Wear*, 265, 741-755.

[16] Gorokhovsky, V., Bowman, C., Gannon, P., Van Vorous, D., Voevodin, A. A., Rut-
      kowski, A., Muratore, C., Smith, R. J., Kayani, A., Gelles, D., Shutthanandan, V., &
      Trusov, B. G. (2006). *Surf. Coat. Technol.*, 201, 3732-3747.

[17] Chang, Chi-Lung, Chen, Jun-Han, Tsai, Pi-Chuen, Ho, Wei-Yu, & Wang, Da-Yung.
      (2008). *Surf. Coat. Technol.*, 203, 619-623.

[18] Park, In-Wook, & Kim, KwangHo. (2002). *Journal of Materials Processing Technology*,
      130-131, 254-259.

[19] Lee, Eung-Ahn, & Kim, KwangHo. (2002). *Thin Solid Films*, 420-421, 371-376.

[20] Ma, S.L., Ma, D.Y., Guo, Y., Xu, B., Wu, G.Z., Xu, K.W., & Chu, Paul K. (2007). *Acta-
      Materialia*, 55, 6350-6355.

[21] Guo, Yan, Ma, Shengli, Xuand, Kewei, & Bell, Tom. (2008). *Nanotechnology*, 19,
      215603.

[22] Kim, Soo Hyun, Kim, Jong Kuk, & Kim, KwangHo. (2002). *Thin Solid Films*, 420-421,
      360-365.

[23] Rebouta, L., Tavares, C. J., Aimo, R., Wang, Z., Pischow, K., Alves, E., Rojas, T. C., & Odriozola, J. A. (2000). *Surf. Coat. Technol.*, 133-134, 234-239.

[24] Khizhnyak, N. A. (1965). *Sov. Phys. Technol. Phys.*, 35, 847.

[25] Aksenov, I. I., Belous, V. A., & Padalka, V. G. (1978). USSR Authors Certificate No. 605425, (Rus.).

[26] Aksenov, I. I., et al. (1978). *Prib. Tekhn. Ehksp.*, 5, 236, (Rus.).

[27] Anders, A., Anders, S., & Brown, I. G. (1995). *Plasma Sources Sci. Technol.*, 4, 1-12.

[28] Martin, P. J., & Bendavid, A. (2001). *Thin Solid Films*, 394, 1-15.

[29] Aksenov, I. I., Vasilyev, V. V., Druz, B., Luchaninov, A. A., Omarov, A. O., & Strel'nitskij, V. E. (2007). *Surf. Coat. Technol.*, 201, 6084-6089.

[30] Byon, E., Kim, J.-K., Kwon, S.-C., & Anders, A. (2004). *IEEE Trans. Plasma Sci.*, 23, 433-439.

[31] Aksenov, I. I., Zaleskij, D.Yu., & Strel'nitskij, V. E. (2000). 1st International Congress on Radiation Physics, High Current Electronics and Modification of Materials, September, Tomsk, Russia. Proceedings 3, 130.

[32] Takikawa, Hirofumi, & Tanoue, Hideto. (2007). *IEEE Trans. Plasma Sci.*, 35(4), 992-999, Aug. 2007.

[33] Karpov, D. A. (1997). *Surf. Coat. Technol.*, 96(1), 22-33, Nov. 1997.

[34] Martin, P. J., Bendavid, A., & Takikawa, H. (1999). *J. Vac. Sci. Technol. A, Vac. Surf. Films*, 17(4), 2351-2359, Jul. 1999.

[35] Anders, A. (1999). *Surf. Coat. Technol.*, 120/121, 319-330.

[36] Martin, P. J., & Bendavid, A. (2001). *Surf. Coat. Technol.*, 142-144, 7-10.

[37] Anders, A. (2002). *Vacuum*, 67(3/4), 673-686, Sep. 2002.

[38] Aksenov, I. I., Strel'nitskij, V. E., Vasilyev, V. V., & Zaleslij, D. Y. (2003). *Surf. Coat. Technol.*, 163/164, 118-127.

[39] Shi, X., Tay, B. K., Tan, H. S., Liu, E., Shi, J., Cheah, L. K., & Jin, X. (1999). *Thin Solid Films*, 345(1), 1-6, May 1999.

[40] Tay, B. K., Zhao, Z. W., & Chua, D. H. C. (2006). *Mater. Sci. Eng. R*, 52(1-3), 1-48, May 2006.

[41] Ryabchikov, A. I., & Stepanov, I. B. (1998). Rev. Sci. Instrum., 69(2), 810-812, Feb. 1998.

[42] Zimmer, O. (2005). *Surf. Coat. Technol.*, 200(1-4), 440-443, Oct. 2005.

[43] Chun, S. Y., Chayaraha, A., Kinomura, A., Tsubouchi, N., Heck, C., Horino, Y., & Fukui, H. (1999). Jpn. J. Appl. Phys., 38(4B), L467-L469, Apr. 1999.

[44] Meyer, C. F. H., & Scheibe, J. (1999). Presented at the Int. Conf. Metallurgical Coatings Thin Films (ICMCTF), San Diego, CA, Paper B4-9.

[45] Taki, Y., Kitagawa, T., & Takaki, O. (1997). *J. Mater. Sci. Lett.*, 16(7), 553-556, Apr. 1997.

[46] Yoshitake, T., Shiraishi, G., & Nagayama, K. (2002). Appl. Surf. Sci., 197/198, 379-383.

[47] Ning Jiang, Y. G., Shen, H. J., Zhang, S. N., & Bao, X. Y. Hou. (2006). *Materials Science and Engineering: B*, 135, 1-9.

[48] Guo, Y., Shengli, M., & Xu, K. (2007). *Surf. Coat. Technol.*, 201, 5240-5243.

[49] Abraham, S., Choi, E. Y., Kang, N., & Kim, K. H. (2007). *Surf. Coat. Technol.*, 202, 915-919.

[50] Guo, Y., Ma, S. L., Xu, K. W., Bell, T., Li, X. Y., & Dong, H. (2008). *Key Engineering Materials*, 373, 188-191.

[51] Qin, C. P., Zheng, Y. G., & Wei, R. (2010). *Surf. Coat. Technol.*, 204, 3530-3538.

[52] Shtansky, D. V., Levashov, E. A., Sheveiko, A. N., & Moore, J. J. (1999). *Metallurgical And Materials Transactions A*, 30, 2439-2447, JCPDS File No: 42-1489.

[53] Diserens, M., Patscheider, J., & Lévy, F. (1998). *Surf. Coat. Technol.*, 108, 241-246.

[54] Phinichka, N., Chandra, R., & Barber, Z. H. (2004). J. Vac. Sci. Technol. A., 22, 477-482.

[55] Johnson, L. J. S., Rogström, L., Johansson, M. P., Odén, M., & Hultman, L. (2010). Thin Solid Films, 519, 1397-1403.

[56] Oliver, W.C., & Pharr, G.M. (1992). *J. Mater. Res.*, 7, 1564-1583.

[57] Dayan, M., Shengli, M., & Xu, K. (2005). *Surf. Coat. Technol.*, 200, 382-386.

[58] Jeon, J. H., Choi, S. R., Chung, W. S., & Kim, K. H. (2004). *Surf. Coat. Technol.*, 188-189, 415-419.

[59] Veprek, S., & Reiprich, S. (1995). *Thin Solid Films*, 268, 64-71.

[60] Xu, H., Nie, X., & Wei, R. (2006). *Surf. Coat. Technol.*, 201, 4236-4241.

[61] Sundgren, J.-E., Johansson, B.-O., Hentzell, H. T. G., & Karlsson, S.-E. (1983). *Thin Solid Films*, 105, 385.

[62] Zerkout, S., Achour, S., & Tabet, N. (2007). *J. Phys. D: Appl. Phys.*, 40, 7508.

[63] Pfeiler, M., Kutschej, K., Penoy, M., Michotte, C., & Mitterer, Kathrein M. (2007). *Surf. Coat. Technol.*, 202, 1050.

[64] Lee, D. N. (1989). *Journal of Material Science*, 24, 4375.

[65] Zhao, J. P., Wang, X., Chen, Z. Y., Yang, S. Q., Shi, T. S., & Liu, X. H. (1997). *J. Phys. D: Appl. Phys*, 30, 5.

[66] Zhang, X., Weber, W. H., Vassell, W. C., Potter, T. J., & Tamor, M. A. (1998). *J. Appl. Phys.*, 83, 2820.

[67] Jiang, N., Shen, Y. G., Zhang, H. J., Bao, S. N., & Hou, X. Y. (2006). *Materials Science and Engineering*, B 135, 1.

[68] Fallon, P. J., Veerasamy, V. S., Davis, C. A., Robertson, J., Amaratunga, G. A., Milne, W. I., & Koskinen, J. (1993). *Phys. Rev. B*, 48, 4777.

[69] Oliver, W. C., & Pharr, G. M. (2004). *Journal of Material Research*, 19, 3.

[70] Chun, Sung-Yong. (2010). *Journal of the Korean Physical Society*, 56, 1134.

Chapter 9

Interfacial Electron Scattering in Nanocomposite Materials: Electrical Measurements to Reveal The Nc-MeN/a-SiN$_x$ Nanostructure in Order to Tune Macroscopic Properties

R. Sanjinés and C. S. Sandu

Additional information is available at the end of the chapter

1. Introduction

The impressive number of outstanding physical, chemical and mechanical properties of transition metal nitrides MeN (Me stands for transition metal Ti, V, Cr, Zr, Mo, Nb, Ta,..) makes then very attractive materials for many industrial applications as protective and decorative coatings [1,2], superconducting nanostructured thin films for single photon detectors [3,4], diffusion barriers in microelectronic devices [5,6], catalytic films [7,8], and also as materials for biomedical applications [9,10]. Depending on the oxidation states of the transition metal, the Me-N system can exhibit a rich variety of stable or metastable crystallographic phases. Thus, the tetragonal Me$_2$N and the cubic fcc structures are preferred for IVB-VA compounds (TiN, VN, ZrN) while for VB and VIB-VA compounds the stable phase is the hexagonal one (NbN, MoN, TaN and WN). In particular, as thin films MeN can be easily integrated in microelectronic devices and commonly used as diffusion barriers in magneto-resistive random access memory, resistors, excellent barrier diffusion against Cu, or as preferred barrier absorber material for EUV mask [4-8].

To further improve the performances and efficiency of MeN functional properties, nano-crystalline or amorphous ternary systems, such as Me-Al-N, Me-Si-N, and other Me-X-N forming highly stable compounds have been also investigated [11-30]. By addition of Al or Si to binary MeN, hardness, thermal stability and chemical inertness of the films can be improved [11-16]. In particular TiSiN, TaSiN, NbSiN and WSiN thin films have been mainly investigated as diffusion barriers and electrodes for phase change random access memory (PRAM) devices. [21-30]. The addition of Si leads to the formation of a nanocomposite (nanocrystallites of MeN + amourphous SiN_x) or a solid solution single phase $Me_{1-x}Si_xN$ material [13-16]. In nanocomposite thin films (nc-MeN/a-SiN_x), crystallite sizes are of the order of few nanometers. The density of point defects (vacancies, interstitials, antisites), the grain size, the grain surfaces, and boundary regions play an increased role on physical properties. The arrangement and the chemical composition of the so-called "amorphous" minority phase (SiN_x) are crucial for electrical and mechanical properties [17-20]. The location, composition and the thickness of the amorphous phase must therefore be known precisely.

Usually these films are deposited by CVD [12, 15] or PVD [11, 14, 16] techniques; among the PVD techniques, magnetron reactive sputtering is often used as a low-temperature film growth technique. The macroscopic properties of these films such as mechanical, optical or electrical strongly depend on chemical composition and nanostructure of the resulting films which are influenced by the deposition parameters such as the substrate temperature, the flux and kinetic energy of impinging atomic and ionic species on the surface of the growing film, and the condensation rate.

The aim of this paper is to give a general overview on the relationship between the electrical and structural properties of binary MeN and nanocomposite nc-MeN/a-SiN_x thin films deposited by reactive magnetron sputtering. In particular we will focus on the possibility to use electrical measurements and electron scattering models to obtain pertinent information concerning the chemical composition, thickness and continuity of the insulating layer covering conducting nanocrystallites in nanocomposite films. It is not the purpose of this paper to develop further the models describing film nanostructure. This has already been extensively covered in much of the cited literature. The limitations of characterization techniques, such as HRTEM, XRD and XPS, in revealing such composite nanostructures, as described by various physical models, motivate us to employ unconventional investigation techniques such as electrical measurements in order to evidence, for example, the continuity of the insulating SiN_x-layer on conducting MeN-crystallites. The case of a special type of nanocomposite materials: nanocrystallites of Phase1 surrounded by a very thin interfacial layer of Phase2, obtained as a result of self-segregation is one of the most difficult to investigate. Instead, the goal of this paper is to discuss the ability of electrical measurements to support such models.

2. Film morphology and nanostructure

Depending on the deposition conditions binary transition metal nitride MeN thin films deposited by reactive magnetron sputtering usually crystallize with strong (111) or (200) preferential orientation and exhibit elongated crystallites in the grow direction [16-20,31] as one can notice from XRD or SEM measurements (Fig. 1). In MeXN (X=Si,Ge,B,Cu..), the addition of X leads to important modification of the films morphology. Thus, as a function of increasing X content (C_X), the average crystallite size, d, in many systems such as Ti-Si-N, Ti-Ge-N, Ti-Sn-N, Nb-Si-N, Zr-Si-N, Ta-Si-N, decreases from tens on nm to about 2 nm [16-20, 32-35]. Whether a ternary single-phase or composite multiphased system is formed depends on the chemical reactivity of the involved atoms and on the deposition conditions. In many case X atoms can substitute metal atoms in the fcc MeN lattice up to a critical concentration (limit of solubility, α_x). The segregation of X atoms on the MeN crystallite surface is mainly responsible for the limitation of their growth. It results in the formation of a nanocomposite material composed of a thin amorphous phase on the MeN crystallite surfaces. Frequently a relationship $d \propto 1/C_X$ is observed in MeXN films (see Fig. 2) suggesting that in this regime the increase in the X content determines a simultaneous increase in the surface-to-volume ratio of the MeN crystallites, which is realized by a subsequent decrease in the average crystallite sizes.

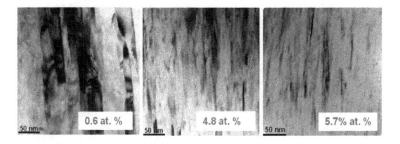

Figure 1. TEM images showing the evolution of the Zr-Si-N films morphology with increasing Si content.

3. Model for the Me-X-N film formation

The sketch given in Fig. 3 illustrates the growth model for the formation of Me-X-N ternary system. As a function of the X content, three X concentration regions can be identified. In the case of PVD deposition techniques such as magnetron sputtering, the film growth is frequently made out of thermodynamical equilibrium. Consequently the addition of X atoms in small quantities into the MeN lattice presents the introduction of structural points defects (substitutions, interstitials, vacancies), which might perturb the crystallite growth. This region 1 is called *Region* 1 or the region of pseudo-solubility of X atoms in MeN. The limit of the pseudo-solubility α_x of X depends on deposition conditions (substrate temperature and bias). Once the X content exceeds α_x the additional X atoms increasingly segregate and accumulate at the grain boundary regions. This concentration region is denoted as *Region* 2, in this region the surface of each X crystallite is progressively coated by a growing XN$_y$ tissue layer up to a certain limit, referred to as the so-called X coverage level, X_{cov}. When X_{cov} =1 (full coverage), further increase of the X content leads to the formation of ultrathin XN$_y$ layer surrounding completely the surface of the MeN crystallite and hindering the crystallite growth. Thus, in the *Region* 3 the, microstructure is strongly altered as a consequence of X segregation.

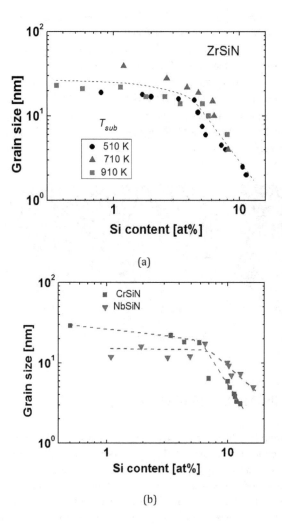

Figure 2. (a)Grain size vs. Si content in Zr-Si-N films. (b) Grain size vs. Si content in Cr-Si-N and Nb-Si-N thin films.

The degree of X surface-coverage of a crystallite of a typical size d can be determined in terms of C_X and C_{Me} concentrations considering a simple model. In a cubic shaped crystallite of volume $V_c = d^3$. For a fcc-NaCl-type structure each unit cell of volume a^3 contain 4 atoms, then the density of Me atoms in V_c is given by $N_{Me/Vc} = (4/a^3)d^3$ while its surface density is $N_{Me/Surf} = (2/a^2)(6d^2)$ (a is the lattice constant). The relation between the number of Me surface atoms and that of the volume is $N_{Me/Surf}/N_{Me/Vc} - (3\,a/d)$. Under the assumption that the segregated X atoms occupy the surface Me sites, the degree of the X coverage (X_{cov}) in terms of C_X and C_{Me} atomic per cent is then given by

Interfacial Electron Scattering in Nanocomposite Materials: Electrical Measurements to Reveal The Nc-
MeN/a-SiN, Nanostructure in Order to Tune Macroscopic Properties

227

$$X_{cov} = \left(\frac{N_{X/Surf}}{N_{Me/Surf}}\right) = \frac{N_{X/Surf}}{N_{Me/Vc}\left(3\frac{a}{d}\right)} = \frac{(C_X - \alpha_L)}{(C_{Me} + \alpha_L)\left(3\frac{a}{d}\right)} \tag{1}$$

In equation (1) the quantity α_L is the limit X solubility and takes into account of the amount
of X atoms that are incorporated in the MeN:Si crystal lattice. As state above, in Me-X-N system the films generally exhibit a pronounced needle-like, columnar structure, with elongated crystallites where the length to width ratio higher than L/d=10 is observed. For such a
situation, the relation (1) can be easily modified by introducing the vertical grain extension L
as an integer multiple of the in plane crystallite dimension $L=nd$, note that in case of cubic-
shaped crystallites n=1. Therefore the Si coverage for elongated crystallite is

$$X_{cov} = \frac{(C_X - \alpha_L)}{(C_{Me} + \alpha_L)\left(2 + \frac{1}{n}\right)\left(\frac{a}{d}\right)} \tag{2}$$

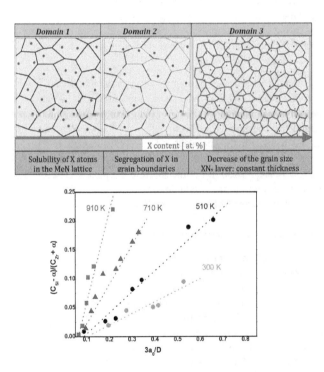

Figure 3. (a) Physical model describing the evolution of nanostructure with increasing X element content. (b) Correlation between secondary phase segregation at the grain boundaries and nanostructure in Zr-Si-N films deposited at
various temperatures.

Interestingly, the relation (1,2) predicts that if X_{cov} remains constant, the average crystallite size d and the X content follow a linear relationship $C_X \approx cte \times \frac{1}{d}$, which is observed in many Me-X-N systems. Fig. 3b for example illustrates that the dependence of $\frac{(C_X - \alpha_L)}{(C_{Me} + \alpha_L)}$ on $\frac{3a}{d}$ for the ZrSiN films is linear and that X_{cov} can be evaluated from the slope of the curve.

4. Electrical properties

The electrical resistivity is strongly dependent on the film nanostructure. Not only the type of polycrystalline major phase and grain boundary phase (metal-like conductor, semiconductor or insulator), but also the grain size of crystalline phase, the thickness of grain boundary phase and the global film density are the main parameters that influence the resistivity of nanocomposite thin films. The thicknesses of the minority grain boundary phase (such as SiN_x, a-C, BN, or $TiGe_y$) can be calculated using the model for the film formation described in the section 3. Due to the fact that the charge carrier scattering is very sensitive to grain size and nature of the grain boundary regions, it should more convenient to plot the d.c. electrical resistivity values as a function of the grain size rather than to consider the atomic concentration C_X of the minority phase.

In the results presented in this section, the reported grain size values were obtained from XRD measurements, most of which were acquired in grazing incidence configuration. This value represents the mean value of the crystallite size in an oblique direction at about 15°-30° with respect to the film normal, the grain size values obtained from grazing incidence XRD are much closer to the lateral size of the crystallites. So, to a first approximation, these values could be considered as more suitable for calculating electrical parameters, due to the fact that the electrical resistivity is measured in the plane of the film. Obviously, some adjustment could be made in order to take into account the real lateral size of the crystallites, which can be obtained from TEM in cross-section.

4.1. Nanostructure and RT d.c. electrical resistivity

Depending on the atomic concentration of the minority phase and on the chemical composition of the main crystalline phase, the room temperature (RT) resistivity of MeXN nanocomposites can change over two or more orders of magnitude. It is worth noting that rather to plot the RT resistivity as a function of the atomic concentration of the minority phase it is more instructive to represent the RT resistivity as a function of the grain size in order to extricate the contribution of the structural film modification on the carriers transport properties. In this section, we will consider the nature of composites, how they can be classified from their dc electrical resistivity behavior, how these reflect the electrical properties of the constituent materials, and, in the next section, to what extent they can be modelled. Depending on the electrical nature of the polycrystalline mayor phase (metal-like conductor or semiconductor) and grain boundary tissue phase (conductor or isolator) three types of nanocomposites will be discussed: metal-like conductor/insulator (M-I), metal-like conductor/conductor (M-M), and semiconductor/insulator (S-I).

4.1.1. Metal-like conductor/Insulator (M-I) interfaces

The room temperature electrical resistivity of Zr-Si-N films, deposited at various temperatures and bias voltages are shown in Fig. 4a [19]. The influence of the crystallite size on the resistivity is clearly observed in the case of films deposited without bias at 510, 710 and 910 K. These films present a nanocomposite structure nc-ZrN/a-SiN$_x$. The formation of an amorphous SiN$_x$ insulating (a-SiN$_x$) layer on the ZrN nanocrystallite (nc-ZN) surface is responsible for significant increases in resistivity only for the films with silicon coverage Si_{cov} greater than 0.5 ML. It should be mentioned that 0.5 ML coverage layer corresponds to 1 ML of SiN$_x$ between two adjacent ZrN crystallites. Wherever such SiN$_x$ layers (thicker than 1.0 ML) are formed, it is observed a significant gap between the resistivity values at the same crystallite size value but at different values for SiN$_x$ thickness. The effect of grain boundary scattering on film resistivity is enhanced as grain size is decreased. This corresponds to the increase of the gap between the resistivity values of films deposited at 510 K, 710 K and 910 K with crystallite size reduction. Thus, the grain boundary scattering is enhanced in the case of the films showing higher SiN$_x$ surface coverage. The increase in resistivity with increasing Si content related to the formation of nanocomposite material showing an insulating and continuum layer between conducting nanocrystallites, has been reported in Zr-Si-N (nc-ZrN/SiN$_x$) [14], Nb-Si-N (nc-NbN/SiN$_x$) [18], Ta-Si-N (nc-TaSiN/SiN$_x$) [36] and Ti-B-N films (nc-TiB/BN) [37].

4.1.2. Conductor/Conductor (M-M) interfaces

In the case of M-M nanocomposites, the presence of a different conducting phase at the grain boundaries of conducting crystallites, does not strongly affect the resistivity behavior. Small changes in the densification, chemical composition of the films and high density of point defects at the grain boundary regions could induce the observed variations. In nitrogen-deficient or nitrogen-rich binary MeN$_{1\pm x}$ thin films, the resistivity can strongly depend on the chemical composition. Thus, in ZrN$_{1\pm x}$ and TaN$_{1\pm x}$ large variations of the resistivity (one to two orders of magnitude) as a function of the N content are observed. The N-deficiency also affects the electrical properties of Me-Si-N nanocomposites. For example, in the case of N-deficient (ZrSi)$_y$N$_x$ (with x≤0.5) films deposited at RT without bias, or at 300 K and 510 K with -150 V bias, it is observed that the resistivity does not change significantly with decreasing grain size (increasing Si content) as shown in Fig. 4a. The Si compositional independent behavior of the resistivity is supposed to be originated from direct percolation of the conducting ZrN$_{1-x}$ crystallites and/or ZrN$_{1-x}$ crystallites separated by low degree of nitridation of the SiN$_x$ grain boundary phase. In fact, the Si_{cov} surface coverage was found to be too small (about 0.3 ML) to completely encapsulate the ZrN crystallites. We have also obtained similar results on N-deficient (TaSi)$_y$N$_x$ nanocomposite films [36]. Resistivity Si compositional independent behavior was also reported for nanocomposite TiN/SiN$_x$ films by Jedrzeovski [38].

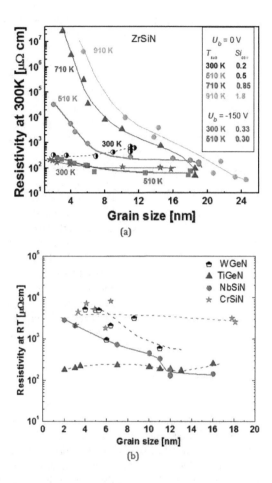

Figure 4. a) Resistivity vs. grain size for Zr-Si-N films deposited at various temperatures and biases. (b) Resistivity vs. grain size for various films.

Furthermore, by comparing the evolution of the resistivity with decreasing grain size for Ti-Ge-N and W-Ge-N composite films, a different behavior is observed (Fig. 4b). This difference gives us information about the electrical nature of the grain boundary phase: conducting $TiGe_x$ phase in the case of Ti-Ge-N films [32] and insulating GeN_x phase in the case of W-Ge-N films [33], which is similar to the insulating SiN_x phase in Nb-Si-N films. In the case of the WC_x-C films, changes in the phase composition from nc-W_2C/nc-WC to nc-WC/a-C are responsible for resistivity variation correlated to the variation of the crystallite size and the presence of high density of point defects [34]. The situation is similar for TiBC films though the presence of three phases, nc-TiB, nc-TiC and a-C, and the large solubility of B in TiC make it difficult the interpretation of results [35]. In WC-C and TiBC nanocomposites, the grain boundary regions composed of a-C do not play a significant role. The main free path of the electrons is mainly limited by the high density of point defects in the amorphous samples whilst lattice defects and grain size predominate in presence of nanocrystalline binary or ternary phases [34,35].

4.1.3. Semiconductor/Insulator (S-I) interfaces

Some MeN such as ScN and CrN are semiconductors. As far as we know, the electrical properties of ScN/SiN$_x$ composites have not been published. In the case of CrN/a-SiN$_x$ system, variation of resistivity with the grain size was also observed [39]. But in the case of a semiconductor material, small variation in the chemical composition of CrN$_x$ crystallites strongly influences the electrical resistivity of the film as shown in Fig. 4b. This could explain the dispersion of the points for the same value of the grain size. This case is the most difficult to model unambiguously. The temperature dependence of the intrinsic resistivity in semiconductor materials masks the temperature dependence of the grain boundary scattering.

4.2. Temperature dependence of d.c. resistivity.

Measuring the electrical resistivity as a function of the temperature gives further information on the main mechanisms responsible of the charge carriers scattering linked to structural changes due to the addition of the second constituent. Fig. 5 shows the temperature dependent d.c. electrical resistivity $\rho(T)$ curves of NbSiN films deposited at 510 K as a function of the Si content [18]. The $\rho(T)$ curves progressively change from metallic-like to nonmetallic-like behavior as the Si content in the films increases. These characteristic trends are often observed in (M-I) type of nanocomposites as a function of the concentration of the insulating minority phase. Fig. 6 a and 6b shows few $\rho(T)$ curves of selected nanocomposite films such as ZrSiN, TiGeN, WC-C, and TiBC for specific grain size. In Fig. 6c are presented $\rho(T)$ curves of $Cr_{0.92}Si_{0.08}N_{1.02}$ and CrN_y for $0.93 \leq y \leq 1.15$. Detailed results concerning temperature dependent electrical resistivity can be found in [18] for NbSiN, in [34, 35] for WC-C and TiBC, and in [19] for ZrSiN.

In the case of (M-I) nanocomposites (Fig. 6a), the temperature dependence of resistivity can easily be correlated with film nanostructure (grain size and thickness of the insulating phase). The effect of the electron scattering at grain boundaries is enhanced by the presence of a thin insulating barrier. Thus, the resistivity $\rho(T)$ of Zr-Si-N films with large grain size exhibits metallic behavior (see [19]) while those having small grains exhibit a negative temperature coefficient of resistivity (TCR $= \dfrac{1}{\rho}\dfrac{\partial \rho}{\partial T}$). Similar behavior was reported by Piloud in the case of TiBN films [40]. In all these works the authors correlates the negative TCR with the diminution of the crystallite size and the presence of an insulating phase between conducting crystallites.

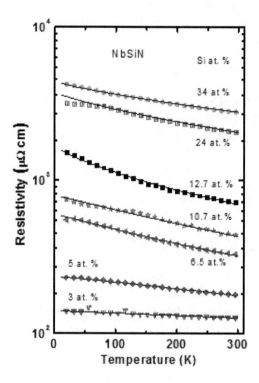

Figure 5. Resistivity vs. Temperature variation for Nb-Si-N films with various Si content.

In the case of (C-C) nanocomposite (TiGeN, WC-C and TiBC), the temperature dependence of resistivity is flat, so the TCR values are low (see Fig. 6b). The resistivity variations for 3 types of films having grain size of about 3nm are similar. The resistivity variation behavior cannot be correlated with the thickness of the phase present at grain boundaries, probably because of a high transmission probability G of charge carriers at the grain boundaries. The absence of the energy gap at GB should be responsible for this.

In the case of (S-I) nanocomposites (CrSiN) the temperature dependence of resistivity (Fig. 6c) cannot easily be correlated with film grain size and scattering probability, because the dependence of polycrystalline semiconducting materials on temperature masks the nano-structure related effects. The change in the N content in CrN_x crystallites significantly influences the resistivity behavior. The resistivity behavior of CrN_x changes from metallic to semiconducting with increasing N content. The formation of a nanocomposite CrN/SiN film with an insulating SiN_x phase between semiconducting CrN crystallites could explain the further increase in resistivity.

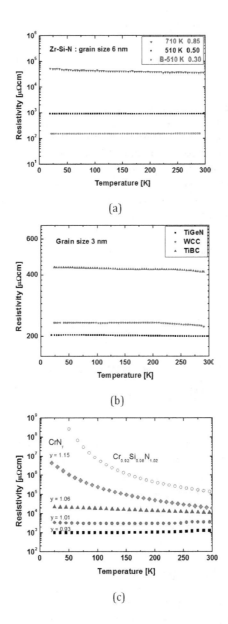

Figure 6. a) Resistivity vs. Temperature variation for Zr-Si-N films with a grain size of 6 nm. (b) Resistivity vs. Temperature variation for various films with a grain size of 3 nm. (c) Resistivity vs. Temperature variation for CrN films with various N/Cr atomic ratios.

5. Grain boundary scattering model

It is frequently observed that the electrical conductivity of thin polycrystalline films strongly deviates from that of the corresponding bulk single-crystalline material. The conductivity is reduced, which commonly is explained by a reduction of the mean free path of electrons (mfp), and often a negative coefficient of resistivity TCR is observed. In the case of quasi-amorphous or heavily distorted materials negative TCR values have been explained by the hopping mechanism or by a week localization of a two-dimensional electron system. However, these models cannot explain all negative TCR values. Based on many experimental results G. Reiss, H. Hoffman et al. [41] proposed the grain boundary scattering model for the d.c. resistivity of polycrystalline thin film materials. The authors state that all electrons reflected by the grain boundaries along one mfp do not contribute to the resulting current and the reduction of the conductivity depends exponentially on the number of grain boundaries per mfp. In this model, an effective mean free path $L_G = L\, G^{(L/D)}$ is introduced to describe the electron scattering at the grain boundaries including the grain size effect; the d.c. electrical conductivity is given by $\sigma = \sigma_B G^{(L/D)}$ where σ_B is the bulk conductivity, G is the probability for an electron to pass a single grain boundary and D is the mean grain size. Under the condition $L/D \ll 1$ the conductivity is reduced to the Drude conductivity without grain boundary effect. The model also predicts a change of the sign of TCR from positive to negative values when L and G fulfil the condition $(L/D)\ln(1/G) > 2$.

Thus, the dc electrical resistivity is then given by

$$\rho_g = \left(\frac{m_e^* v_F}{N e^2} \right) \left(\frac{1}{L} \right) G^{-(L/D)} = \left(\frac{K}{L} \right) G^{-(L/D)} \tag{3}$$

where m_e^* is the effective masse of the charge carriers, v_F is the Fermi velocity, N is the density of the charge carriers, D is the grain size parameter, L is the inner-crystalline mean free path and G is the mean probability for electrons to pass a single grain boundary. The inner-crystalline mean free path L, describing the volume scattering of electrons, is limited by a temperature invariant elastic scattering at lattice defects and acoustic phonons, namely l_e, and by the temperature dependent inelastic scattering, l_{in}, $L^{-1} = l_e^{-1} + l_{in}^{-1}$. The inelastic mean free path is approximated by $l_{in} \approx \alpha T^{-p}$ where α and p are material specific constants.

In nanocomposite materials composed of a main polycrystalline phase (TiN, ZrN, NbN, TaN, CrN, WC, etc) and amorphous minority or tissue phase (SiN$_x$, GeN, TiB, a-C, etc.) the grain size of the main material, and the thickness and nature of the grain boundary regions can be easily tailored by the volume concentration of the minority phase. The equation (3) gives the possibility, by a simple fitting procedure of $\rho(T)$ curves, to obtain pertinent information on the main scattering parameters such as G, D and l_e. For the theoretical modeling via the relation (3), at the first approximation grain sizes obtained from XRD or HRTEM measurements can be used. Regarding the factor $K = \left(\dfrac{m_e^* v_F}{N e^2}\right)$, the Fermi velocity and the electron density N are typically in the range of $v_F \approx 1.0\ 10^8$ cm s^{-1} and $N = (4\text{-}10)\ 10^{22}$ cm^{-3}. More precise N values can be obtained from Hall effect or from optical measurements for stoichiometric or defective MeN$_x$ and MeC$_x$. During the fitting procedure, these values can be adjusted to obtain the best fits.

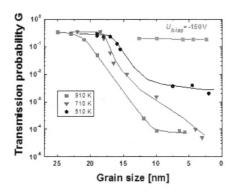

Figure 7. Mean probability G vs grain size for ZrSiN films deposited at 510, 710 910 K and under bias (lines are added to aid the eye).

The transport mechanisms in nc-NbN/a-SiN$_x$, nc-ZrN/a-SiN$_x$ and nc-TaN/a-SiN$_x$ have been satisfactory described by the grain boundary scattering model. In the case of the Zr-Si-N system, the electrical properties of Zr$_z$Si$_y$N$_x$ films deposited various temperatures, namely, at 300 K (without substrate heating), 510 K, 710 K and 910 K have been investigated in details. It is important to point out that by increasing the substrate temperature the solubility limit of Si, α_L, in the ZrN lattice decreases whereas the Si coverage, Si_{cov}, increases. Thus, the values of the pairs (α_L, Si_{cov}) are (5 at. %, 0.2), (4 at. %, 0.5), (2 at. %, 0.8) and (1 at. %, 1.8) for the Zr$_z$Si$_y$N$_x$ films deposited at 300 K, 510 K, 710 K and 910 K, respectively. The main probability for electrons to pass the grain boundary G is related to the formation of the SiN$_x$ coverage layer as shown in Fig. 7. In films deposited at 710 K and 910 K, pure ZrN and Zr$_z$Si$_y$N$_x$ films with low Si content (<0.5 at. %) exhibit high G values, G=0.25-0.35. But, for Si content > 1 at%, where the solubility limit is low and the Si coverage important, G deeply decreases down to small values in good correlation with the high thickness values of the SiN$_x$ grain boundary layer (1.6 ML and 3.6 ML, respectively) in these films. For films deposited at 510 K, G decreases slowly and at higher Si content (> 3 at. %) in good agreement with the higher Si solubility and lower thicknesses of the SiN$_x$ layer observed in these films. The electron transmission probability coefficient, G gives us information concerning the continuity and thickness of the insulating phase between conducting grains. In the case of nanocomposites with SiN_x covering layers thinner than 1.0 ML (300 K ZrSiN and 510K ZrSiN with -150 V bias), G is larger than 0.05. So, a small scattering probability at grain boundaries implies a small barrier at grain boundaries or the percolation of ZrN crystallites. The effect of the nitrogen content on the electrical nature of the SiN_x grain boundary layer has been investigated in ZrSiN (deposited at 510 K and 710 K with -150 V bias) and in TaSiN films (deposited at 653 K). For N-deficient $(ZrSi)_yN_x$ and $(TaSi)_yN_x$ nanocomposites, the transmission probability G remains in the range of 0.1-0.2 over the full investigated Si compositional range(0 – 12 at. %). These results clearly indicate that Si segregation in N-deficient MeSiN films does not lead to the formation of an effective electrically insulating SiN_x layer.

6. SiN_x thickness and resistivity

6.1. Tunneling effect

When two metallic electrodes are separated by an insulating layer (M-I-M structure) the action of the insulating layer is to introduce a potential barrier Φ between the electrodes inhibiting the flow of electrons. However, if the insulating layer is sufficiently thin the current can flow through the insulating region by tunnel effect [42,43]. In the case of electron tunnelling experiments the tunnelling probability is found to be exponentially dependent on the potential barrier width, the tunnelling current is $I_T \propto e^{-\sqrt{\Phi}d} \approx e^{-2.4d}$ and the tunnelling conductance can change by about one order of magnitude for the change $\Delta d \approx 0.1$ nm. Fig. 8 was constructed by considering the thickness of the SiN_x covering layer, as calculated by using the 3-step model for the film formation in the case of ZrSiN films, and the measured resistivity values taken in the region where we have a nanocomposite ZrN/SiN_x structure, as far for the grain size of 4, 6, 8 and 10 nm. The resistivity tends to increases exponentially with the thickness of the SiN_x layer in the range 1.0-3.6 ML (corresponding to a separation distance of 0.2-0.8 nm between metallic crystallites) suggesting that the transport of the electrons across the thin barrier layer seems to occur by tunnelling.

For a M-I-M structure with an insulating layer of thickness d, the tunnelling probability T_p for a with a rectangular barrier with an effective barrier height $e\phi_B$ is given by:

$$T_p = exp\left(-2\left[\frac{2m_e^* e\phi_B}{\hbar^2}\right]^{1/2} d\right) \approx exp\left(-\alpha_T \sqrt{\phi_B}d\right) \tag{4}$$

If the effective masse in the insulator is $m_e^* \approx m_e$, the ϕ_B en volts and d in Å then $\alpha_T=1$. The tunnelling conductivity σ_T is given by

$$\sigma_T = \varepsilon_0 \omega_D^2 \tau_T = \left(\frac{Ne^2}{m_e^* v_F}\right) l_e T_p \tag{5}$$

where τ_T is the tunneling relaxation time $\tau_T = \dfrac{l_e T_P}{v_F}$ and l_e the effective main free path. The tunneling conductivity decreases exponetially with increasing the thickness of the insulating layer. Fig. 9 shows the relationship of the tunneling resistivity and the thickness of the insulating layer with the tunneling probability as calculated from Eq's (4) and (5) for the ZrSiN system. T_P and σ_T have been calculated for two different electron densities $N = (1.8-3.6)\ 10^{22}$ cm^{-3} for the ZrN, $v_F = 10^8$ cm s^{-1}, $l_e = (5-10)$ nm and for two different effective barrier height values, $\phi_B = 0.6$ V and $\phi_B = 1$ V. Considering that 1 ML of SiN$_x$ corresponds to about 0.22 nm, the tunneling model predicts that for $\phi_B = 1$ V the tunneling probability decreases from 10^{-1} to 10^{-4} and the resistivity increases from about 10^2 $\mu\Omega$ cm up to 10^5 $\mu\Omega$ cm when the thickness of the SiN$_x$ layer change from 1 ML to 4 ML. For lower ϕ_B values, equivalents insulating layers lead to low resistivity values. In Fig. 7 it is shown the transmission probability G, obtained by fitting the $\rho(T)$ experimental curves using the grain boundary scattering model, as a function of the crystallite size (deduced from XRD) for the ZrSiN films deposited at various temperatures. It is worth noting that for films exhibiting comparable crystallite sizes, for instace 12 nm, but with different Si coverages, G values are in the rage of 10^{-1}, 10^{-2} and 10^{-3} corresponding to SiN$_x$ thicknesses of 1ML, 1,6 ML and 3.6 ML, respectively. Though the tunneling conductivity in nanopolycrystallite materials is undoubtedly complexe, the correlation between T_P and G is remarkable. These trends suggest that tunneling conduction should be envolved as one of the the conduction mechanisms responsible for electrons to cross de grain boundary layer between two adjacent crystallites; in particular in the case of elongated crystallites where the length to width ratio higher than 10 have been reported from HRTEM investigations for MoN/SiNx nanocomposites [31,32].

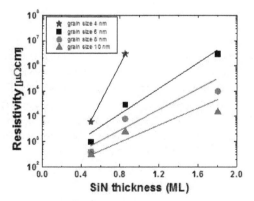

Figure 8. Resistivity vs. thickness of SiN$_x$ interfacial layer: ZrSiN films with 4, 6, 8, and 10 nm crystallite size at 510 K (0.5 ML), 710 K (0.85 ML) and 910 K (1.8 ML). (lines are added to aid the eye).

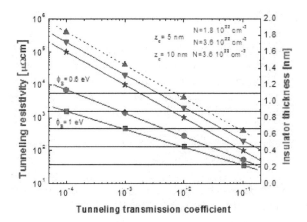

Figure 9. Tunneling resistivity and interfacial insulating thickness vs. tunneling coefficient.

It will be useful to estimate the barrier height $e\phi_B$ in such M-I-M structures. Knowing this value, the transmission probability across metallic-insulating-metallic structures can be calculated as a function of SiN_x thickness. By determining G from fitting the resistivity dependence on temperature, we can extract the SiN_x thickness from electrical measurements. We will not speculate that the calculated G values are sufficiently precise to then extract the energy gap at the grain boundaries. Rather, we would just like to highlight the good correlation between structural and electrical properties.

6.2. I-V characterisation

To investigate if the observed conductivity in SiN_x thin film results from tunnelling of electrons through the SiN_x thin film current-voltage (I-V) measurements should be performed on Me-SiNx-Me structures. For this purpose, SiN_x films sandwiched between ZrN or TaN have been prepared by magnetron sputtering. These structures have been deposited at 740 K and with bias voltage of -150 V in order to obtain smooth surfaces leading to relatively sharp interfaces. The I-V characteristics for $ZrN/SiN_x/ZrN$ structures with different SiN_x thicknesses are show in Fig. 10. The effect of the SiN_x thickness is clearly noticed by comparing the I-V curves with that of the structure without SiN_x layer. For SiN_x thicknesses small than 5 nm (namely the ultrathin regime) the I-V curves show ohmic behaviour, while for thicknesses higher or equal to 5 nm the I-V curves exhibit a symmetric non-linear behaviour. Similar curves have been also observed in the case of $TaN/SiNx/TaN$ structures. The linear behaviour observed in the ultrathin regime can be interpreted in terms of electron tunneling process. A Poole-Frenkel type resistance describes the S-shaped curves, often observed in thin films. In the case of ideal symmetric M-I-M structure the tunnelling current for $V < \phi_B$ is given by [42]

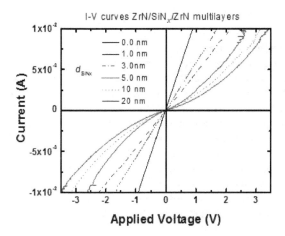

Figure 10. I-V curves in ZrN/SiN/ZrN multilayer films with various insulating SiN layer thicknesses.

$$I = I_0 \left[(\phi_B - V/2)exp(-A\sqrt{(\phi_B - V/2)}) - (\phi_B + V/2)exp(-A\sqrt{(\phi_B + V/2)}) \right] \qquad (6)$$

and for low V range

$$I = \frac{(2m\phi_B)^{1/2}e^2}{2h^2 d} V exp\left[-2d\sqrt{\frac{2me\phi_B}{\hbar^2}} \right] \qquad (7)$$

Earlier studies of the current transport mechanisms in silicon nitride thin films, performed on structures such as $Au/Si_3N_4/Mo$ and $Au/Si_3N_4/Si$, have shown that the current transport is essentially independent of the substrate material, the film thickness and the polarity of the electrodes [44]. In these studies the Si_3N_4 thickness was in the range of 30 to 300 nm. Depending on the ambient temperature and the electric field three different conduction mechanisms have been identified: Ohmic-type, Poole-Frenkel emission and Fowler-Nordheim tunneling. The Poole-Frenkel mechanism is mainly due to field-assisted excitation from traps and is often observed on defective materials while the Fowler-Nordhein conduction depends on free carriers tunnelling through high quality Si_3N_4 at high electric fields. Ohmic conduction was attributed to the hopping of thermally excited electrons from one isolated state to another.

$$\text{Poole-Frenkel} \qquad I_{PF} = C_{PF} V exp(-e\phi_B + aV^{1/2}/kT) \tag{8}$$

$$\text{Fowler-Nordhein} \qquad I_{FN} = C_{FN} V^2 exp(-b/V) \tag{9}$$

$$\text{Ohmic-type} \qquad I_{Om} = C_{Om} V exp(-e\phi_O /kT) \tag{10}$$

Tao et al [45] have been investigated the effect of N vacancies (Si-Si bonds) and O substitutions (Si-O bonds) on the current-transport properties of $SiN_{1.06}$, $SiN_{1.33}$ and $SiO_{1.67}N_{0.22}$ thin films. The thickness of the Si nitride and of the Si oxynitride layers in Al/SiNO/Si/In structures was typically 15 nm. The results of these studies have been well correlated with the nature of the insulating layer. Thus, all the films exhibit an Ohmic regimen at low electrical fields. The ohmic resistivity depends on the nature of the film; Si-rich films exhibit lower resistivity values while oxynitrides films show the highest values, as the carriers are generated by thermal excitation from traps it was concluded that the density of traps is higher in Si-rich films than in oxynitrides. At intermediate and high electrical fields, Poole-Frenkel emission is the dominant conduction mechanism in Si-rich SiN_x films whereas Fowler-Nordhein tunnelling is mainly involved in oxynitrides films but absent in Si-rich films. Both Poole-Frenkel (at intermediate electrical fields) and Fowler-Nordhein (at high fields) mechanisms are present in nearly stoichiometric Si_3N_4 films.

Based on all these studies we can conclude that the tunnelling current-transport in ultrathin SiN_x layers is very sensitive to N vacancies and to the presence of oxygen atoms. Therefore, Nc-MeN/a-SiN_x nanocomposite thin films containing silicon nitride layers with similar thicknesses but with different chemical composition (sub-stoichiometric or nearly stoichiometric Si_3N_4, oxynitride) can exhibit different electrical properties. Thus, the effects of N-deficiency on the electrical properties of ZrN/SiN_x and TaN/SiN_x nanocomposites as discussed in the section 4 can be interpreted in terms of the presence of high density of free carriers at the grain boundaries thereby leading to high tunnel currents. In addition, the difficulty with real interfaces in thin films is that even if the chemical composition were well controlled surface roughness would increase the local electrical field rising up unexpectedly the tunnel currents.

7. Conclusion

Nanocomposite materials present a high degree of complexity due to small grain size, high curvature radius of nanocrystallites and, in general, a very thin minority phase layer situated at the grain boundaries. Correlating electrical resistivity measurements with film nanostructure provides information concerning the thickness and continuity of the interfacial layer covering conducting nanocrystallites in conducting-insulating nanocomposite films. Aside from some constraints, the possibility to measure experimentally, albeit indirectly, such small interfacial layer thicknesses constitutes an important breakthrough in precise characterization of such nanostructures.

Acknowledgements

The authors wish to thank the Swiss National Science Foundation and the EPFL for financial support.

Author details

R. Sanjinés* and C. S. Sandu

*Address all correspondence to: rosendo.sanjines@epfl.ch

EPFL-SB-ICPM-LPMC, Ecole Polytechnique Fédérale de Lausanne, CH-1015 Lausanne, Switzerland

References

[1] Toth, . E. (1971). *Transition Metal Carbides and Nitrides*, Academic, New York.

[2] Holleck, H., & , J. (1986). *Vac. Sci. Technol. A*, 4, 2661.

[3] Kaloyeros, A. E., & Eisenbraun, E. (2000). *Annu. Rev. Mater. Sci.*, 30, 363.

[4] Riekkinen, T., Molarius, J., Laurila, T., Nurmela, A., Suni, I., & Kivilauhti, J. K. (2002). *Microelectron. Eng.*, 64, 289.

[5] Rossnagel, S. M. (2002). *J. Vac. Sci. Technol.*, B20, 2328.

[6] Wittmer, M. (1980). *Applied Physics Letters*, 36, 456.

[7] Daughton, J. M. (1992). *Thin Solid Films,*, 216, 162.

[8] Sun, X., Kolawa, E., Chen, J., Reid, J. S., & Nicolet, M. A. (1993). *Thin Solid Films*, 236, 347.

[9] Leng, Y. X., et al. (2001). Thin Solid Films, , 398-399.

[10] Wallrapp, F., & Fromherz, P. (2006). *J. Appl. Phys.*, 99, 114103.

[11] Diserens, M., Patscheider, J., & Lévy, F. (1998). *Surf. Coat. Technol.*, 108-109.

[12] Veprek, S., et al. (1999). *J. Vac. Sci. Technol.*, A 17, 2401.

[13] Musil, J. (2000). *Surf. Coat. Technol.*, 125, 322.

[14] Pilloud, D., Pierson, J. F., Marques, A. P., & Cavaleiro, A. (2004). *Surf. Coat. Technol.*, 180-181.

[15] Veprek, S., Maritza, J. G., & Veprek-Heijman, . (2008). *Surf. Coat. Technol.*, 202, 5063.

[16] Sandu, C. S., Sanjinés, R., Benkahoul, M., Medjani, F., & Lévy, F. (2006). *Surf. Coat. Technol.*, 201, 4083.

[17] Martinez, E., Sanjinés, R., Banakh, O., & Lévy, F. (2004). *Thin Solid Films*, 447-448.

[18] Sanjinés, R., Benkahoul, M., Sandu, C. S., Schmid, P. E., & Lévy, F. (2005). *J. Appl. Phys.*, 98, 123511.

[19] Sandu, C. S., Medjani, F., & Sanjinés, R. (2007). *Rev. Adv. Mater. Sci.*, 15-173.

[20] Sandu, C. S., Harada, S., Sanjinés, R., & Cavaleiro, A. (2010). *Surf. Coat. Technol.*, 204, 1907.

[21] Reid, J. S., Kolawa, E., Ruiz, R. P., & Nicolet, M. A. (1993). *Thin Solid Films*, 236, 319.

[22] Kim, D. J., Kim, Y. T., & Park, J. W. (1997). *J. Appl. Phys.*,, 82, 4847.

[23] Lee, Y. J., Suh, B. S., Kwom, M. S., & Park, C. O. (1999). *J. Appl. Phys.*, 85, 1927.

[24] Suh, Y. S., Heuss, G. P., & Misra, V. (2002). *Appl. Phys. Lett.*, 80, 1403.

[25] Letendu, F., Hugon, M. C., Agius, B., Vickridge, I., Berthier, C., & Lameille, J. M. (2006). *Thin Solid Films*, 513, 118.

[26] Olowolafe, J. O., Rau, I., Mr, K., Unruh, C. P., Swann, Z. S., Jawad, T., & Alford, . (2000). *Thin Solid Films*, 365, 19.

[27] Hübner, R., Hecker, M., Mattern, N., Hoffmann, V., Wetzig, K., Heuer, H., Wenzel , Ch , Engelmann, H. J., Gehre, D., & Zschech, E. (2006). *Thin Solid Films*, 500, 259.

[28] Cabral, C., Jr, Saenger, K. L., Kotecki, D. E., & Harper, J. M. E. (2000). *J. Mater. Res.*, 15, 194.

[29] Alén, P., Aaltonen, T., Ritala, M., Leskelä, M., Sajavaara, T., Keinonen, J., Hooker, J. C., & Maes, J. W. (2004). *J. Electrochem. Soc.*,, 151, G523.

[30] Jung, K. M., Jung, M. S., Kim, Y. B., & Choi, K. D. (2009). *Thin Solid Films*,, 517, 3837.

[31] Sandu, C. S., Sanjinés, R., & Medjani, F. (2008). *Surf Coat. Technol*, 202, 2278.

[32] Sandu, C. S., Sanjinés, R., Benkahoul, M., Parlinska-Wojtan, M., Karimi, A., & Lévy, F. (2006). Thin Solid Films ., 496, 336.

[33] Piedade, A. P., Gomes, M. J., Pierson, J. F., & Cavaleiro, A. (2006). Surf. Coat. Technol ., 200, 6303.

[34] Abad, M. D., Sánchez-López, J. C., Cusnir, N., & Sanjinés, R. (2009). *Journal of Applied Physics*, 105, 033510.

[35] Abad, M. D., Sanjinés, R., Endrino, J. L., Gago, R., Andersson, J., & Sánchez-López, J. C. (2011). *Plasma Process and Polymers*, 8, 579.

[36] Oezer, D., Ramirez, G., Rodil, S. E., & Sanjinés, R. submitted to J.A.P

[37] Pierson, J. F., Bertran, F., Bauer, J. P., Jolly, J., & Surf, Ž. (2001). *Coat.Technol.*, 142-144.

[38] Jedrzejowski, P., Baloukas, B., Klemberg-Sapieha, J. E., & Martinu, L. (2004). *J. Vac. Sci. Technol. A*, 22, 725.

[39] Martinez, E., Sanjines, R., Banakh, O., & Levy, F. (2004). *Thin Solid Films*, 447-448.

[40] Pilloud, D., Pierson, J. F., & Pichon, L. (2006). *Materials Science and Engineering B*, 131, 36.

[41] Reiss, G., Vancea, J., & Hoffman, H. (1986). *PhysRev.Lett*, 56, 2100.

[42] Simmons, J. G. (1963). *J. Appl. Phys.*, 34, 1793.

[43] Fisher, J. C., & Giaever, I. (1961). *J. Appl. Phys.*, 32172.

[44] Sze, S. M. (1967). *J. Appl. Phys.*, 38, 2951.

[45] Tao, M., Park, D., Mohammad, S. N., Li, D., Botchkerav, A. E., & Morkoç, H. (1996). *Phil. Mag. B,*, 73, 723.

Permissions

The contributors of this book come from diverse backgrounds, making this book a truly international effort. This book will bring forth new frontiers with its revolutionizing research information and detailed analysis of the nascent developments around the world.

We would like to thank Dr. Farzad Ebrahimi, for lending his expertise to make the book truly unique. He has played a crucial role in the development of this book. Without his invaluable contribution this book wouldn't have been possible. He has made vital efforts to compile up to date information on the varied aspects of this subject to make this book a valuable addition to the collection of many professionals and students.

This book was conceptualized with the vision of imparting up-to-date information and advanced data in this field. To ensure the same, a matchless editorial board was set up. Every individual on the board went through rigorous rounds of assessment to prove their worth. After which they invested a large part of their time researching and compiling the most relevant data for our readers. Conferences and sessions were held from time to time between the editorial board and the contributing authors to present the data in the most comprehensible form. The editorial team has worked tirelessly to provide valuable and valid information to help people across the globe.

Every chapter published in this book has been scrutinized by our experts. Their significance has been extensively debated. The topics covered herein carry significant findings which will fuel the growth of the discipline. They may even be implemented as practical applications or may be referred to as a beginning point for another development. Chapters in this book were first published by InTech; hereby published with permission under the Creative Commons Attribution License or equivalent.

The editorial board has been involved in producing this book since its inception. They have spent rigorous hours researching and exploring the diverse topics which have resulted in the successful publishing of this book. They have passed on their knowledge of decades through this book. To expedite this challenging task, the publisher supported the team at every step. A small team of assistant editors was also appointed to further simplify the editing procedure and attain best results for the readers.

Our editorial team has been hand-picked from every corner of the world. Their multi-ethnicity adds dynamic inputs to the discussions which result in innovative outcomes. These outcomes are then further discussed with the researchers and contributors who give their valuable feedback and opinion regarding the same. The feedback is then collaborated with the researches and they are edited in a comprehensive manner to aid the understanding of the subject.

Apart from the editorial board, the designing team has also invested a significant amount of their time in understanding the subject and creating the most relevant covers. They scrutinized every image to scout for the most suitable representation of the subject and create an appropriate cover for the book.

The publishing team has been involved in this book since its early stages. They were actively engaged in every process, be it collecting the data, connecting with the contributors or procuring relevant information. The team has been an ardent support to the editorial, designing and production team. Their endless efforts to recruit the best for this project, has resulted in the accomplishment of this book. They are a veteran in the field of academics and their pool of knowledge is as vast as their experience in printing. Their expertise and guidance has proved useful at every step. Their uncompromising quality standards have made this book an exceptional effort. Their encouragement from time to time has been an inspiration for everyone.

The publisher and the editorial board hope that this book will prove to be a valuable piece of knowledge for researchers, students, practitioners and scholars across the globe.

List of Contributors

Bahman Nasiri–Tabrizi, Abbas Fahami and Reza Ebrahimi–Kahrizsangi
Materials Engineering Department, Najafabad Branch, Islamic Azad University, Najafabad, Isfahan, Iran

Farzad Ebrahimi
Department of Mechanical Engineering, Faculty of Engineering, Imam Khomeini International University, Qazvin, Iran

Hema Bhandari, S. Anoop Kumar and S. K. Dhawan
CSIR–National Physical Laboratory, India

Masoud Mozafari, Mehrnoush Mehraien and Lobat Tayebi
Helmerich Advanced Technology Research Center, School of Material Science and Engineering, Oklahoma State University, USA

Daryoosh Vashaee
Helmerich Advanced Technology Research Center, School of Electrical and Computer Engineering, Oklahoma State University, USA

Dongfang Yang
National Research Council Canada, 800 Collip Circle, London, Ontario, Canada

Vladimir Dzyuba, Yurii Kulchin and Valentin Milichko
Institute of Automation and Control Processes of Russian Academy of Science, Vladivostok, Russia
Far Eastern Federal University, Vladivostok, Russia

Jow-Lay Huang and Pramoda K. Nayak
National Cheng Kung University, Taiwan

Nelcy Della Santina Mohallem and Victor L. Guimarães
ICEx – Universidade Federal de Minas Gerais, Brazil

Juliana Batista Silva
Centro de Desenvolvimento de Tecnologia Nuclear - CNEN/CDTN, Brazil

Gabriel L. Tacchi Nascimento
Nanum Nanotechnology Ltda, Brazil

Seunghun Lee, P. Vijai Bharathy, T. Elangovan, Do-Geun Kim and Jong-Kuk Kim
Korea Institute of Materials Science, Changwon, Republic of Korea

R. Sanjinés and C. S. Sandu
EPFL-SB-ICPM-LPMC, Ecole Polytechnique Fédérale de Lausanne, CH-1015 Lausanne, Switzerland

Printed in the USA
CPSIA information can be obtained
at www.ICGtesting.com
JSHW011435221024
72173JS00004B/810

9 781632 383976